PRENTICE HALL
Biophysics and Bioengineering Series
Abraham Noordergraaf, Series Editor

Regulation and Control Mechanisms in Biological Systems

Vishnampet S. Vaidhyanathan

State University of New York at Buffalo

P T R Prentice Hall, Englewood Cliffs, New Jersey, 07632

Library of Congress Cataloging-in-Publication Data

Vaidhyanathan, Vishnampet S.
    Regulation and control mechanisms in biological systems /
Vishnampet S. Vaidhyanathan.
        p.   cm.
    Includes bibliographical references and index.
    ISBN 0-13-771262-6
    1. Biological control systems. 2. Thermodynamics. I. Title.
QH508.V35 1993
574.1'88--dc20   92-8695
                                                             CIP

Editorial/production supervision
  and interior design: *Brendan M. Stewart*
Prepress buyer: *Mary Elizabeth McCartney*
Manufacturing buyer: *Susan Brunke*
Acquisition editor: *Betty Sun*

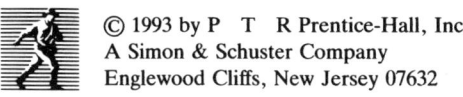 © 1993 by P T R Prentice-Hall, Inc.
A Simon & Schuster Company
Englewood Cliffs, New Jersey 07632

The publisher offers discounts on this book when ordered in bulk quantities. For more information, write: Special Sales/Professional Marketing, Prentice Hall, Professional Technical Reference Division, Englewood Cliffs, NJ 07632.

Printed in the United States of America
10 9 8 7 6 5 4 3 2 1

ISBN 0-13-771262-6

Prentice-Hall International (UK) Limited, *London*
Prentice-Hall of Australia Pty. Limited, *Sydney*
Prentice-Hall Canada Inc., *Toronto*
Prentice-Hall Hispanoamericana, S.A., *Mexico*
Prentice-Hall of India Private Limited, *New Delhi*
Prentice-Hall of Japan, Inc., *Tokyo*
Simon & Schuster Asia Pte. Ltd., *Singapore*
Editora Prentice-Hall do Brasil, Ltda., *Rio de Janeiro*

Contents

Chapter 4   *Aspects of System Behavior in Stationary States*   93

Chapter 5   *Biochemical Control Mechanisms*   122

Chapter 6 *Mathematical Aspects*   162

Chapter 7 *Some Control Mechanism*
            Models   198

Chapter 8 *Nonlinear Nonequilibrium*
            Thermodynamics   220

Preface

The subject matter of this book, namely, regulation and control in biology, is very extensive and diverse. A unified and complete review of various aspects of the subject will form a very lengthy text. Evidently it is not possible to consider all information available on both descriptive and quantitative aspects of regulation in biology. In addition, the time has not yet arrived in this subject when a comprehensive treatment of the field is both feasible and useful.

Certain theoretical aspects of physics, chemistry, and applied mathematics relevant to our understanding of various biological control mechanisms form the structure of this book. Many important aspects of homeostatic control mechanisms are not presented for want of space and for the additional reason that excellent descriptive coverage of subjects not presented here are available in many other books. These omitted subjects include compartmental models, optimization methods, and analysis of physiological signals. Therefore, this book is not intended to be a complete and rigorous account of various theoretical models or of dynamic systems.

Nature exhibits a phenomenal variety of regulation and control mechanisms. This includes mimicking of the flower of a species of orchids in Australia to resemble the body of a female wasp, so that the male wasp, in its effort to copulate, will help to pollinate the flowers. Similarly, sage bushes shed their large leaves in dry season to preserve their water content and also secrete an oil which is unpleasant to goats, thereby saving the bushes from being consumed. This is also an example of regulation and control in biology.

The integrative action of biological systems as contributing to their ability to adapt successfully to environmental demands has been known for some time. Recent advances in aspects of control in biological systems are mostly biochemical and accounts of these are descriptive. The homeostatic mechanisms can be understood on the basis of our knowledge of physics and chemistry, and may be quantified if the detailed chemical reactions are known. Science is concerned with certain observed facts and reasoning provided by the observation. Systems change their properties or parameters with time. An objective in physics is to determine the state of the system or its variables at a later specified time on the basis of the information available at a specified initial time.

It was difficult to decide to whom and at what level a book of this subject should be addressed. Many of my colleagues admitted candidly that they would be reluctant to undertake writing a book on this subject. Having agreed to write this monograph, the realization of the difficulty involved in selecting materials to be included soon seeped in. Evidently no arrangement is going to be satisfactory to everyone. A descriptive approach would be redundant, since many other excellent books are available. A mathematical approach should not duplicate the books of Riggs, or Milsum, or Segel. However, I feel a book emphasizing the relevant quantitative aspects containing information about nonequilibrium situations may be useful. With this in view, subjects like hormonal regulation, kidney physiology, or thermal regulation are not presented. Even subjects like optimality or cardiac functions are not included. I have included only those subjects which I feel are informative to the student in order that he or she can fully appreciate control mechanisms devised by nature and attempts by humans to understand them.

I wish to express my deep appreciation of the cooperation and help accorded to me by most of my colleagues, specifically Drs. Anbar, Loonsk, Slaughter, and Spangler. Professor Fred Snell and Mr. Tony Dasaro have been helpful with their critical comments. Many other students have contributed to the final draft of this book.

Finally, I wish to dedicate this book to my professor, Dr. Scott E. Wood who, in his own philosophical manner, taught me more about life than about the sciences.

—*V. S. Vaidhyanathan*

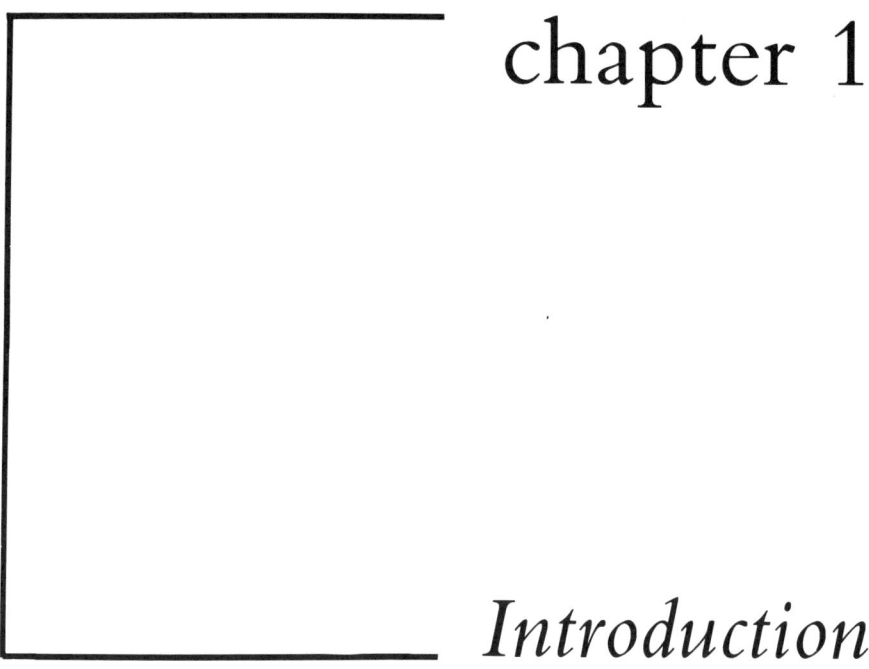

chapter 1

Introduction

PRELIMINARY REMARKS

Organisms acquire freedom from their environment by means of homeo-static mechanisms. Biology is engrained in the physical and chemical processes, as well as communication with the system's external environment, through material and energy transport. Theoretical biology attempts to construct the basic principles of biology to describe the behavior of biological systems. It is a basic tenet of theoretical biology that life processes are explicable in terms of our knowledge of physics and chemistry. In this book a description of the characteristics of biological regulatory systems with major emphasis on physical and chemical aspects is provided. Regulation implies that certain changes brought about externally to the system are minimized.

The concentrations in cells of various ions and other important species are regulated by means of homeostatic processes. The need for regulation arises from the imbalance between supply and demand and reflects a change in the intake, in the use, or in the excretion of materials from cells. Thus, concentration of a specific species can be maintained in a compartment with connections with other compartments and the external environment by either production or consumption of this species by chemical reaction or by transport from outside, resulting in either intake or in the outflow of this species. The conservation relations of matter and energy and the basic principles of chemistry govern the chemical reactions and related processes that occur within the cell. Thus, flow processes and biochemical control mechanisms form the significant aspects of regulation and control in living systems. The internal temperature within the body is also regulated in homeotherms. Recent developments in nonequilibrium thermodynamics and dissipative structure theory suggest that the energy generated by the cell is utilized to maintain the system in a far-from-equilibrium state, constituting the driving force of the life processes. Thus, energy balance, its production and dissipation, as well as energy absorption are important in the study of homeostasis.

In order to gain an understanding of the mechanisms which govern the functions of living systems, it is essential to unravel the physical and chemical principles working at the intracellular, subcellular, interorganismic, and molecular levels. It is becoming increasingly clear that organization is a fundamental component of biological systems. Biological systems are extremely complex and complicated. Complexity is an inherently occurring state of living systems. In a multicellular organism, there are approximately ten billion cellular units operating in a functionally coordinated manner. Even a single cell is built up in an enormously complex manner. In single bacterial cells, more than ten million biological macromolecules interact coherently so as to sustain the state of material order, which we regard as life.

Philosophers and scientists have pondered over the nature of life—the essential property that a living organism has, but that which a dead organism has lost—for many centuries. Modern science has come to the conclusion that life consists simply of the possession of specific intact molecular structures within and between cells. As long as these structures are present, life persists. Once they break down, life no longer exists.

Certain very small animals defy conventional definitions of life. These odd animals, called nematodes, are able to enter a state of suspended animation, persisting for years, even decades, in a death-like state, only to revive within minutes in the right conditions. They go into this suspended state by drying up, shriveling into wrinkled husks of their former selves. Sometimes they became so dry that they crumble when touched. In the dry form they can last for years without eating or breathing. Digestion and mobility are stopped. Nervous systems are inert and no metabolism occurs. However, a few drops of water on these little creatures will bring back life, and the nematodes swell back to normal proportions and crawl away, hungrily looking for food. The fact that some animals can do this implies that the usual definitions of life cannot be correct. The usual definitions are the criteria such as the ability to grow, reproduce, and carry out metabolism. One is forced to conclude that these animals can die and resurrect.

Thus, life is possibly not a process but a set of interlocked molecular structures with the potential to perform metabolic processes in the presence of water. To enter the state of cryptobiosis, these animals must dry up very slowly. The animal shrinks, but no vital structures crack or separate. The vital structures appear to fold and curl in an orderly manner. It is assumed that possibly trehalose and glycerol replace water on the surfaces of every structure within and between cells during drying, forming a protective coating. So long as the structural integrity of the organism is intact, it is alive. Given the right conditions, the conventional aspects of life are renewed.

Regarded in this manner, one begins to wonder whether the ideas and methods of the contemporary physical sciences are sufficiently advanced to deal with systems of such complexity. The great variety of biological systems does not encourage the formation of a global answer. In connection with the question of the origin of life, one concludes that the actual transition from nonliving matter to a living system is attributed to the phase of molecular self-organization.

In ecology, it is presumed that where many different kinds of plants and animals live together, there will be a better balance than when there are only a few kinds. Thus, it is implied that complexity leads to stability. A measure of species diversity should also be a measure of system diversity. Systems that function through an array of intersecting pathways that provide alternative channels for flow of information or energy do become more stable, the more crossroads there are. Colinvaux (1978) presents the thesis that stability and balance are not so much functions of life acting on life as they are only reflections of the underlying stability of physical systems. He questions the validity of ecological thought claiming that stability is a function of biological complexity.

Living systems require an overall strategy for regulation due to the multivariable and highly interconnected organization. Organisms are self-regulating, adaptive systems capable of autoduplication, acquiring energy from the environment for utilization in their biological functions. But to be self-regulating and adaptive, control and information are essential. Cybernetic mechanisms must therefore provide the basis for the organism's living characteristics as well as the adaptability required for its survival.

In the last three decades following publications of many books on regulation and control aspects of biology, much experimental information has become available. The consensus is that life seems to employ any useful mechanism available to control and optimize its survival and propagation. Dynamic models can be useful in predicting the long-term behavior of physical processes, especially when it is not possible to look into the actual system itself. Feedback controls can modify the performance. With proper choice of controls, one can modify and improve the performance of a system. A set of conditions that must be met in order to insure optimality should be specified.

Levels of control within the cell should be defined and identified. Very little is known about the basic mechanism of control of cell division rate. The interrelationship between the cell environment and the proliferative behavior of cells is one of the major fields of study in homeostatic mechanisms. The manner of functioning of a regulatory mechanism in the context of a particular enzyme reaction or a metabolic sequence should be understood.

Any consideration of metabolic regulation must focus on those enzymes that exercise control over the velocity (flux) of reaction pathways. Fluxes of reaction pathways are properties of the whole system and cannot be adequately examined one step at a time. It requires investigation into the quantitative interrelations between component parts. Three factors have been identified as crucial to control (Kacser and Burns, 1973). These are controllability, sensitivity, and elasticity. Controllability defines the response of the enzyme in isolation to an inhibitor or perturbation. Sensitivity measures whether a small change in enzyme concentration brings about any change in the overall flux of the pathway. Elasticity is the percentage change in the rate brought about by a change in concentration of the substrate, product, or the effector. Evaluation of these three parameters and the components that regulate them can permit an understanding of the factors which regulate a pathway under a given set of circumstances. Since metabolic reactions may serve multiple functions, modulation of these reactions by a single parameter would not provide the organism with the flexibility required.

The ability of an enzyme to respond to concentrations of metabolites other than its substrate and product adds a new dimension to metabolic regulation. It allows the end product of the metabolic pathway to bring about feedback inhibition in earlier steps. This kind of regulation in strategically located enzymes can have profound effects on cellular metabolism. The concentrations of the various molecules necessary for the functioning of vital life processes are regulated by mechanisms which account for the consumption or the synthesis of these species. Changes in body temperature in homeotherms serve to indicate corresponding changes in the balance of energy by either production or absorption and dissipation. Within the framework of general physicochemical analysis, the flow of energy through an organism is found to be channeled through the synthesis, storage, and utilization of high-energy phosphate bonds.

Organisms, however, never become completely independent of their environment. Despite the fact that living systems are open systems in the thermodynamic sense, and as such continually exchange matter and energy with their surroundings, they have acquired the ability to maintain a constant internal environment.

Homeostasis is truly dynamic in nature, despite its proper distinction from the input-output servo type of control. The physiological stationary state is neither just a passive resistance to change, nor a mere compliance to patterns imposed from the external world. It results mainly from compensatory adjustments actively programmed within the organism in response to the total relevant information it has available.

CHARACTERISTICS OF LIVING SYSTEMS

1. Living organisms are composed of both large and small molecules which are built up from atoms. Thus, the living organisms are physical objects which should follow the laws of physics and chemistry. Living organisms possess the properties of programmed development, differentiation, growth, and movement.

2. An organism is viewed as an organized collection of many physicochemical processes. The viability and behavior exhibited by the organism are a direct consequence of the manner in which these component processes are organized and interrelated in a nontrivial manner.

3. The biological function of a living system is a highly organized assembly of chemical reactions in which various elementary reacting processes have both space and time correlations. A generalized systematic formulation for chemical reaction kinetics in order to analyze the time evolution of these kind of processes is needed. The physicochemical principles involved in the subcellular phenomena at the molecular level should be understood to appreciate the functioning of cells and organisms.

4. Living systems are relatively stable. Thus, stability theory is important in discussing physical theories of life. The intracellular environment of cells must be kept relatively constant. When changes in the environment occur, such as in pH, temperature, ionic composition, or substrate and product concentrations, the intracellular function of the cell must be altered in order to adjust to the changing conditions and enable the maintenance of the internal environment in its optimum state. Body temperature is a typical result of a balance between energy input and energy output.

5. Living systems are open systems in contact with their surroundings. An organism is continually exchanging material with its environment. The cells have to synthesize each of the constituents required at a rate required for harmonious growth, without overproduction or underproduction.

6. It is suspected that stability and order are achieved by living systems by maintenance of their state in a state far from equilibrium. The thermodynamic theory of stationary states related to open systems has applications of considerable significance to living phenomena.

7. At each level of biological organization, homeostatic control mechanisms, mediated through a variety of regulatory subsystems, maintain the organism in a dynamic equilibrium within the relatively narrow confines of the conditions consistent with healthy viability. An essential feature of these mechanisms is that they enable the cell or the whole organism to adapt to changes in both internal and external environmental conditions. A necessary means of explaining some aspects of regulation and control processes can be obtained from studies of systems theory.

8. Physiology of an entire organism cannot be vaguely described by component processes, such as diffusion, mass transport, and protein synthesis. When an inanimate system is isolated or placed in uniform surroundings, an evolution of the system occurs during which all gradients in thermodynamic parameters are leveled out, all permissible reactions occur, and all other forms of energy become degraded into internal energy. Ultimately, the state of equilibrium prevails, and the system has reached the state of maximum entropy (maximum disorder).

9. The laws of thermodynamics require that if a system is not in equilibrium and any process occurs, its entropy increases and disorder manifests. When stationary states are maintained through the influence of constrained forces and fluxes, the rate of such entropy production is a constant, independent of time. The main contribution to entropy production in living systems arises from metabolism. Metabolism comprises processes by which assimilated food is broken down into simpler substances with the release of energy that is required for vital functions.

10. Schrodinger suggested that a living system succeeds in keeping alive, or postpones the attainment of the final equilibrium state of death, by withdrawal of negative entropy from the environment. One of the basic functions of an organism is to rid the system of the entropy, which it cannot avoid producing while being kept alive.

In the course of its history, physics has frequently stepped outside the area of direct experience. A physical problem is intractable sometimes, not due to inadequacy or inability of the accepted concepts of theory, but rather on account of the complexity of the phenomenon under investigation. The prime example of a complex system is a living system. The great majority of chemical reactions that occur within the cell are enzymatically catalyzed. Nonlinear phenomena find wide applications in the analysis of biological phenomena. Examples can be found in the origin of self-oscillations, the formation and propagation of solitary waves (solitons), spontaneous origination of the dissipative structures, as well as in the phase transitions in thermodynamically nonequilibrium systems. Such phenomena are termed bifurcations. The very existence of living beings and their development form a chain of bifurcations in biological structures and processes.

Two trends may be highlighted in contemporary biological sciences. The first trend is connected with extensive applications of the exact sciences. This interdisciplinary approach has resulted in cooperative endeavors between biologists and physicists. Physicists believe that the fundamental laws of the material world are the same in living and nonliving nature and that there exist no specific vital laws. A second trend has the opposite philosophy. That is, an opinion popular among biologists, though not expressed explicitly, is that biological properties rule out the possibility of describing living objects within the framework of physical theories. The formation of this trend was partly stimulated by the reckless application of physical theories to biological phenomena, which as a rule leads to absurdity. Concepts such as pain and motivation are difficult to formulate in terms of purely physical language. Organized behavior of organisms is definitely beyond the reach of the mathematical and physical sciences.

It is an indispensable condition that the success of control depends on recognition. The recognition of a wanted substance among others of a similar nature is essential for the success of biological regulation. The aspects of living systems presented so far describe the superficial nature of the questions to which we seek answers.

The operational reason for the cybernetic mechanism's existence is its stabilization and conservation of the ergonic component. Thus, adaptivity, control, survival, and stability are all different facets of the same biological characteristic.

Upon closer examination, one observes that regulation and control do not necessarily lead to constant internal conditions. It becomes obvious that life processes consist of many rhythmic and oscillatory phenomena. Metabolic and activity patterns exhibit so-called circadian rhythms. We now present elements of the analysis of physiological signals based on systems theory.

CONTROL ANALYSIS BASED ON SYSTEMS THEORY

Systems analysis may be defined quite generally as the application of organized analytical modeling techniques appropriate to explaining complex multivariable systems, many of whose functional components may be initially quite imperfectly measured or even largely unidentified. The elements which should be controlled in a biological system may be defined as follows. *These must be variables which are essential to the continuing stability of the system. At the same time, these variables may be forced into suboptimal or more critically intolerable values by changes in the environment or in other organismal parameters.*

In this sense, the parametric changes are the *forcing functions,* and the resulting alterations in the essential variable are the responses. Both dynamic and stationary-state responses differ specifically not only in their relation to the input but also in their consequence to the organism. To be coupled effectively into a controlled system, the required controlling device must have appropriate sensitivity, speed of action, precision, and stability.

The study of physiological systems is usually approached in terms of cause-effect relationships, which are manifested as the stimulus-response relations. The functional identification of the system is accomplished by such experiments, in the absence of detailed information about the inner structure of the physiological systems. The observed data on physiological systems may be classified broadly as (1) stationary processes and (2) nonstationary processes. The heartbeat, or the shape of the cardiac pressure output, or the electrocardiogram, do not change during short intervals of time. These are called the stationary processes. On the other hand, the electrical activity of the brain is an example of a nonstationary biological output. It is possible that the study of any one of the regulatory mechanisms will be plagued by the simultaneous operation of all the other regulatory mechanisms which are not specifically under study. The theory of control engineering is based on modeling of the physical systems by differential equations. The linear dynamic analysis is generally tractable and yields often comprehensible results. Our presentation of a qualitative perspective of the physiological systems analysis can only be brief.

The regulatory features of the chemical reactions involving the enzymes, as in metabolic regulation, do not exhibit readily the principles involved in the physical control systems. The question arises as to whether the metabolic reaction systems can be adequately represented by feedback diagrams whose algebraic solutions correspond to the known solutions of the metabolic systems. Many different diagrams can yield the same result. The same set of relations may be represented by several diagrams, which are not necessarily simple transformations of one another. Feedback diagrams of chemical systems represent the regulatory aspects of those systems in a way which frequently gives a false picture of the underlying mechanisms. It is also true that feedback diagrams are not always misleading representations of the chemical control systems (Reiner, 1968).

A controlled system is an arrangement of physical components connected or arranged in such a way as to command, direct, or regulate itself on another system. Input is the stimulus or excitation applied to a controlled system from an external source, usually to produce a specific response from the system under control. Output is the actual response from the controlled system, and the response may or may not be proportional to the input. The relation between the input and output quantities is described mathematically by the transfer function.

In an uncontrolled system, the mode of operation is invariable. In a controlled system, the mode of operation is variable. The input in a controlled system is information from the system obtained by the detector and fed back into the input stage of the system. The output is determined by the *effector* acting on the system. *The essential key to control is the occurrence of the feedback, which permits the control system to compare what the effector is doing with what it is supposed to be doing.* In a living system, examples of the effectors are muscle cells of various kinds, cells which secrete various chemicals into the bloodstream, or other extracellular fluids, or cells which transport substances from one part of the body to another. In a simple system, the control element subtracts the input from the feedback, and the sign of this difference is opposite to the sign of the regulated change in the output. Hence, this is called the *negative feedback*. If the feedback is instead positive, the system will usually become unstable, since positive feedback amplifies the input signal to a larger value.

The basic method of modeling is to write equations, which are usually differential equations, to describe the input–output quantities. The art of modeling consists of reducing these equations to the least complexity while still accounting for the significant phenomena with which the investigator is concerned. The system identification results in the determination of the system transfer characteristics without specifying its internal topological structures. For simple systems, the circuit diagram is a convenient method of summarizing the interconnections between individual elements and the direction of processes, when stored energy is exchanged between the elements. Different forms of energy inputs and interconnections between components and diagrams of signal flow are commonly summarized in a block diagram.

Block diagrams should be viewed as a means of representing the functional relations between several variables in a mathematical or graphical form. Since these mathematical relations may be written in many different ways, the block diagram may bear little resemblance to the physical system even though it depicts the appropriate mathematical relations. The blocks appearing in a given diagram may be viewed as a physical system by which input and output variables are related. These relations may be obtained from underlying physical laws, or in quite a few instances by purely empirical observations.

In contrast with this physical point of view, one may also regard block diagrams in a purely mathematical sense. In this situation, the functions appearing in each block are transformations of one variable into another. The algebraic expressions are the rules by which one can calculate the output from the input. When such equations are solved for a specific variable, in terms of the other variables, the constants present in such algebraic relations are called the *system parameters*. The system parameters can assume both positive and negative values. A linear system can be fully described either in the *time domain* or in the *frequency domain*. All information concerning the time response of the system is contained in the frequency characteristics of the system.

The flow graph presented in Figure 1.1 represents the three compartments of a system, together with all the flows. Each compartment is represented by a circle. Conservation laws are satisfied in each of these compartments, which result in certain known relations between several flows. The interconnections between the compartments are shown by heavy lines to signify material flows between compartments. These flows may be

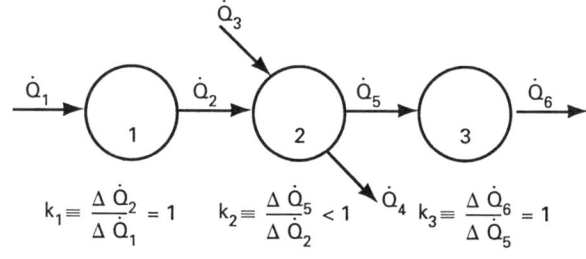

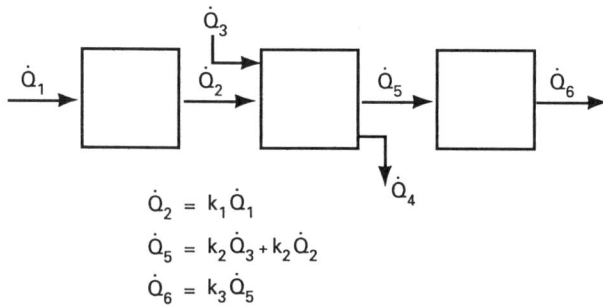

$$\dot{Q}_2 = k_1 \dot{Q}_1$$

$$\dot{Q}_5 = k_2 \dot{Q}_3 + k_2 \dot{Q}_2$$

$$\dot{Q}_6 = k_3 \dot{Q}_5$$

Figure 1.1 Flow graph and block diagram for a three compartment system. The superdots indicate time derivatives. $\dot{Q}_i$ are the flows. Relations exhibited are derived with the relations,

$$\dot{Q}_1 = \dot{Q}_2$$

$$\dot{Q}_2 + \dot{Q}_3 = \dot{Q}_4 + \dot{Q}_5$$

$$\dot{Q}_5 = \dot{Q}_6$$

[Reproduced with permission from Jones].

regarded as signals. The flow graph is reasonably isomorphic to the physical system it represents, in that most of the physical variables can be identified on the graph, whereas this is usually not the case with block diagrams.

The block diagram, also shown in Figure 1.1, provides another way of representing this system in which not only the flows but any other physical quantities serving as the input and output variables can be depicted. The relations between the variables can be given either an analytical equation form or a graphical picture. The block diagram treats all variables as signals and a conservation law, if applicable, is buried within the defining relations. As a result, the output signal from a block may be directed to any number of other blocks without regard for conservation of the signal. Flow diagrams, on the other hand, represent physical compartments and must be drawn such that input and output signals are conserved.

A system may be defined as a collection of components arranged and interconnected in a definite manner. The components may be physical, chemical, biological, or a combination of all three. A system may contain a collection of communicating materials and processes, which together perform some function. The system's behavior is determined by the characteristics of the components of the system, and the structure and communication channels between the components or between the subsystems. The state of the system is defined by a set of numerical values which the variables have at any specified instant of time. The input variables are to some extent affected by the output feedback mechanisms.

Dynamic systems vary their behavior with time, as a result of forces acting upon the system. Generally, systems have internal energy-storing and energy-dissipating components. The term *black box* is used to define a part of the system performing a specific function in which there is one input and one output. There is no other information available concerning the and processes giving rise to specific operation from this component. Sometimes the investigator has some of the information about some of these system components. In this situation, it is called a *grey box*. In a grey box, one has partial information concerning the structure and process which realize the input–output relation.

In controlled systems, the actual output is determined by inputs, which in turn are affected by the outputs, by feedback paths. Control mechanisms attempt to minimize the error, *e*. This is usually accomplished with negative feedback, by the so-called *loop action*. An amplifier is a device by which the level of feedback signal is increased, in either amplitude or in power. The amplification is achieved since the input is able to modulate the supply of, for instance, the electrical energy.

In automatic cameras, the aperture opening or time of exposure is automatically controlled by a mechanism which regulates the incident light. The amount of light received by the photographic film is controlled more easily if the exposure time is calculated in advance. Thus, in this control system, there is a light meter which receives the available light and computes in advance the duration of time that the lens should be opened. This is an example of an open-loop control. Open-loop control has the advantage of being an economic way of control, but its disadvantage is that it is not self-checking. This kind of compromise achieved is also found in certain biological systems.

Biological controlling mechanisms have certain distinct features not usually found in man-made systems. There are significant distinctions not only in the organization, but also in the manner in which the two systems operate. Systems analysis is based on the premise that if the stimulus and response are known or are measured variables, it should be possible to estimate the properties of the system. In the study of the system's behavior, one equates an understanding with a knowledge of how the components contribute to the performance of the whole system. Even for a simple biological process, such a study of the components and their characteristics is a very cumbersome process. Man-made systems can be analyzed mathematically and their behavior described as analytical functions of changes in various parameters. From such analysis, one can predict the limitations of the control system, and the conditions under which the system parameters can lead to instability of the system.

When an input into a system is changed in any manner, the output will also change but not generally in a precisely similar manner. When there is a delay in the transmission of signals at one or more locations along the transmission path, the output is likely to oscillate around the stationary-state value. Delays are present in almost all living control systems. Delays can sometimes be exponential. The transfer function of an exponential delay states that *the rate of change of the response in any one moment is proportional to the difference between the stimulus and response at that moment.* Such delays occur in the hormonal links between receptors and effectors. In living systems, the effectors can work only in one direction. A muscle, for example, can only push so that our limbs are moved in one direction by one set of muscles and moved in the opposite direction by another set of muscles. Similarly, there exist both excitatory and inhibitory hormones.

The human body regulates the temperature, the glucose concentration in the blood, the cardiac output, and other bodily functions effectively by obtaining information from the system by feedback loops. However complicated the control system may be, in order to effect control of a specific property of the system, there should be a mechanism for measurement and a feedback path for this information to the controlling component.

A feedback loop exists if a variable x determines the value of another variable y and the variable y in turn determines the value of the variable x. The variable x could be identified as the input and y is identified as the response (output) of the system. The process of obtaining information about the road conditions by the driver of an automobile, which enables him or her to steer the car correctly, can be regarded as an example of negative feedback. If the output variable y becomes more than a desired value, the feedback loop enables one to reduce the output. One may also have positive feedback. Persons with high blood pressure may have damaged their blood vessels which reduces the pressure in their kidneys. This information requires the central nervous system (CNS) to increase blood pressure, which accelerates the rate of damage. This in turn leads to still higher blood pressure.

Control systems may be classified into various types. The kind of control present in a regulated system is important, since it affects the efficiency of control and the speed with which the system is restored to the desired normal value after a perturbation. Five different types of controls are known to exist. These are (1) directed control, (2) proportional control, (3) proportional plus derivative control, (4) integral control, and (5) control by anticipation of future events.

Briefly, the on-and-off action of a heater in a room is an example of directed control. This kind of control is obviously crude. Addition of a cooling system to the heating system will lead to a more efficient control of the room temperature at a desired level. If the thermostat with a heating current proportional to the deviation of the temperature from the desired value is utilized, obviously a better control which is quicker and more efficient will be obtained. This is an example of a proportional control.

The constant of proportionality is known as the *stiffness* of the system. In a system controlling the flow of heat or water, the load is likely to increase with an increase in the rate of flow, for example, as a result of viscous friction in the pipes. Thus, there will be an error present under steady-state conditions. This error increases with an increase in the velocity of the output element, and therefore is known as the *steady-state velocity error*.

The steady-state error can be minimized, but it can never be reduced to zero. Reduction of the steady-state error is of interest in connection with living systems. In the case of man-made control systems, this can be achieved in two ways. The first one is to introduce a memory element, which activates the controller in proportion to all the errors of the past, so as to reduce errors of the future. The second method of reducing such errors is to feed into the controller a signal which is proportional to the rate of input into the system. This is known as the *input feedforward*.

Derivative control is the use of a feedback signal of a term determined by the rate of change of the error. This procedure has the advantage of correcting for an anticipated overshoot of the correction, from the rate of the decrease of the error. The value of the derivative control is not limited to correction of the overshoot by the control system. If the derivative control has been employed, it will come into play whenever the error is rising, with the net result that feedback processes start to work on anticipated errors, so that the actual errors are smaller than they would otherwise have been.

Integral controls are based on the utilization of a term proportional to the integral over time of the error, added to the feedback signal. The longer the error persists, the greater is the value of the integral, and thus the larger is the compensating correction effort, even though the error has become small. Integral control does not seem to be used widely in living systems. The existence of temporal summation in the nervous system appears to be a kind of integral control. The glucose concentration regulation in the blood is also suspected to act by such integral control mechanisms.

One could envision the possibility of further improving the derivative and integral controls in man-made systems by the addition of a feedback signal, proportional to higher-order derivatives and repeated integrals. Something similar to this does occur in biological systems. When knowledge of past event history is utilized to change the system characteristic so as to give an optimum operation under expected situations, it is called adaptation in biological systems (Reiner, 1968).

Control systems can sometimes get out of order. Excessive feedback is likely to be as serious a handicap to organized activity as defective feedback. When the response to the feedback signal by the system results in an overshoot, it is possible for the system to go into uncontrollable oscillations. In faulty systems, the feedback can become positive instead of being negative, and the system will have a larger error. The error may increase vastly and quickly, thus making the system unstable. Arrythmia in the human body, as well as certain pharmacological agents utilized to control heartbeats, are examples of this situation. Chapter 9 of this book is devoted mainly to such oscillations, and chaotic states as well as catastrophe situations are presented briefly later in this book. The extent to which life is entirely "free" and "independent" of its environment will depend on the precision of the control systems involved.

REDUCTION OF ERRORS BY THE USE OF FEEDBACK

An elementary analysis of the correction effect of an error by use of negative feedback follows, assuming instantaneous response by the system. The signal received from the error-sensing device, shown in Figure 1.2, is fed back and added to the input signal. The reduction of the error achieved by this device is easily computed as follows.

The input signal S_1 into the control system causes the response of the system into the production of an output signal S_2. If the desired output signal is denoted by S_d, the error is $S_e = \{S_2 - S_d\}$. This signal is fed back into the system, summed with the signal S_f. Thus, the new input signal becomes $S_1^\star = \{S_f + S_i\}$.

The gain G equals the ratio $[S_2/S_1]$, and this is called the amplification factor. The feedback factor $[S_f/S_e]$ is denoted by B. For negative feedback systems, B is by definition negative. It is evident that the error S_{e0} that

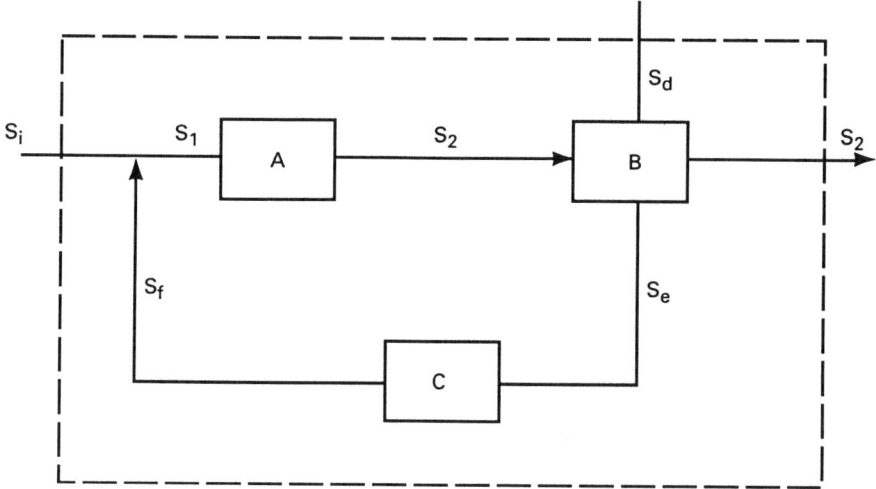

Figure 1.2 A control system, in which the feedback signal is added to the input. A denotes control signal transducer of gain G. B denotes error sensing device. C denotes feedback transdudcer.

would be present in the absence of feedback, is given by

$$S_{e0} = \{GS_i - S_d\}$$
$$S_f = BS_e = B[S_2 - S_d]$$
$$= B\{GS_1 - S_d\}$$
$$= BG\{S_i + S_f\} - S_d$$

(1.1)

Therefore,

$$S_f\{1 - BG\} = B\{GS_i - S_d\} = BS_{e0}$$
$$S_e = S_2 - S_d = GS_1 - S_d$$
$$= G\{S_i + S_f\} - S_d$$
$$= GS_f + S_{e0}$$
$$= [\{BGS_{e0}\}/\{1 - BG\}] + S_{e0}$$
$$= S_{e0}\{1/[1 - BG]\}$$

(1.2)

For negative feedback, B is negative and thus the coefficient of S_{e0} is less than unity. Therefore, the feedback causes the error to be multiplied by a factor less than unity. Consequently, negative feedback causes the error produced to be less than what it was before.

A linear system is one in which the response is proportional to the amount of input (stimulus). Thus, in a linear electric feedback system, a negative error is in all respects similar to the positive error except for the change in the sign. The change in the sign in this situation indicates merely a change in the direction of current flow. Therefore, the same feedback mechanisms which correct a positive error in such a system would similarly correct a negative error. This ability to handle both positive and negative errors with the same mechanism is not a universal property of feedback control systems.

In biological systems, the mechanisms utilized to correct positive errors are often different from the mechanisms utilized to correct negative errors. In such cases, one mechanism may be called the *antagonist* of the other. Generally, it is easier to add another solute to solutions, but it is extremely difficult to remove a particular solute. Similarly in biochemical systems, one often finds it convenient to add an antagonist of a feedback. The secretion of adrenalin during times of stress so that the heart pumps more blood more efficiently is an example of this mechanism. In poststress periods, the human body secretes other chemicals to slow down the heart rate.

When feedback is possible and stable, its advantage is to make performance less dependent on the load. The feedback serves to diminish the dependence of the system on the characteristics of the servomechanism, and serves to stabilize it for all frequencies below a threshold value. The effect of a large negative feedback, if at all stable, will be to increase the stability of the system for low frequencies, but generally at the expense of its stability for some high frequencies. The oscillations caused by an excessive amount of feedback raise the question about the frequencies of the incipient oscillation. The analysis of a controlled system can be accomplished by the use of known properties of the component elements to set up the appropriate differential equation relating the output to the input. The time response of the system for the input signal can then be obtained by solving the resultant differential equation for that particular input. Such analysis, however, is very often possible only for linear systems.

The dynamic behavior of the systems depends on the time–domain response to various inputs such as the initial conditions or disturbances. One generally assumes that each variable is a constant with respect to position variables and varies only with time. Real systems are not homogeneous and, therefore, the system variables are functions of both time and spatial coordinates. The mathematics of considering such dependencies on both space and time are the realm of partial differential equations. For simplicity, many biological system servomechanisms are analyzed usually as ordinary differential equations. This procedure is called *lumping*. A lumped system may be comprised of several different discrete units. Each discrete unit is a lump having uniform homogeneous value of its describing variables at any specified time.

STEP RESPONSE OF A FIRST-ORDER SYSTEM

The response of a linear first–order system to a step input may be described by the differential equation of the following kind:

$$T_1\{dN_o/dt\} + N_o = N_i \qquad (1.3)$$

where N_i is the input of some appropriate quantity, such as an applied voltage, and N_o is the output. T_1 is a time constant of the system related to certain parameters of the system. This relation is usually evaluated from physical description.

A sudden change in the output N_i applied at time t equal to 0 is called a step function input. The response of the first-order system for such a step function change is

$$N_o = N_i\{1 - e^{[-t/T_i]}\} \qquad (1.4)$$

The plot of the ratio $\{N_o/N_i\}$ versus $\{t/T_i\}$ yields the normalized step response of the first-order system. This result is shown in Figure 1.3.

STEP RESPONSE OF A SECOND-ORDER SYSTEM

A second-order control system may similarly be described by a second-order differential equation of the following kind:

$$K\{d^2W/dt^2\} + LK\{dW/dt\} = [\{Mk_1k_2\}/R][W_i - W_o] \qquad (1.5)$$

where W_o is the output and W_i is the input. K is a constant parameter of the system. L is a coefficient of friction, and the second term in the left-hand side of Equation 1.5 depicts the frictional force. Equation 1.5 with W_o equal to

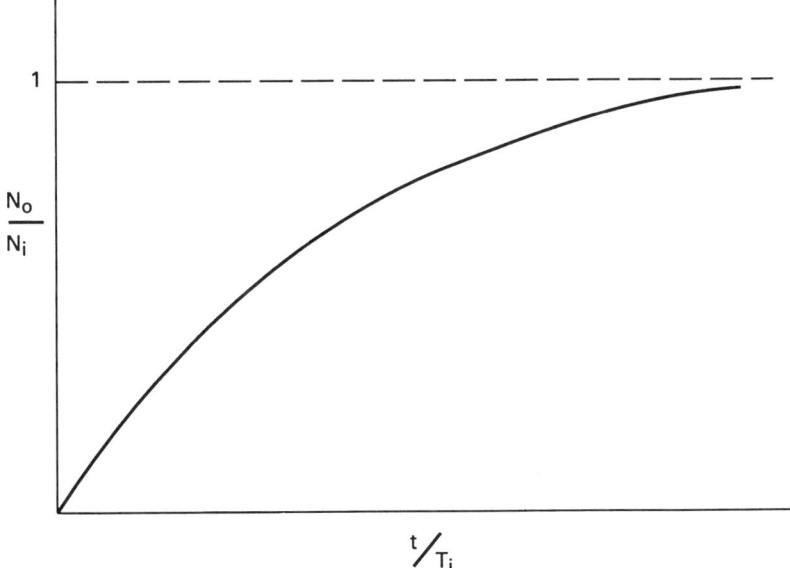

Figure 1.3 Exponential response of a first-order system, to a step input.

zero describes the differential equation satisfied by a simple pendulum undergoing oscillations while being immersed in a viscous fluid.

The step response of a second-order system is obtained by solving the equation:

$$\{d^2W/dt^2\} + 2Fw_n^2\{dW/dt\} + w_n^2W_o = w_n^2W \tag{1.6}$$

where W is the magnitude of the step input. w_n is called the natural frequency of the system. The solution of Equation 1.6 may be written as

$$W_o = W + Ae^{(p_1t)} + Be^{(p_2t)} \tag{1.7}$$

where A and B are constants, and p_1 and p_2 are the roots of the quadratic equations (called characteristic roots or the eigenvalues),

$$p^2 + 2Fw_n p + w_n^2 = 0$$
$$p_1, p_2 = w_n\{-F + [F^2 - 1]^{(1/2)}\} \tag{1.8}$$

The solution can take several forms depending on the magnitudes of the eigenvalues. It is evident that the magnitude and signs of the eigenvalues depend on the frictional (damping) coefficient. The response of a system to a given forcing depends upon the nature of the roots, whether they are: (1) real and unequal ($F > 1$), (2) real and equal ($F = 1$), (3) complex ($0 < F < 1$), or (4) they are imaginary ($F = 0$).

If F equals zero, the roots are purely imaginary and the solution is:

$$W_o = W[1 - \cos(w_n t)] \tag{1.9}$$

The response of the system in this case is purely oscillatory. This is shown by the plot of Figure 1.4a.

The three possible cases for a system with a positive frictional coefficient, $F > 0$, are

1. If the value of frictional coefficient F is greater than zero but less than unity, $0 < F < 1$, the roots of the characteristic equation are complex. The solutions in this case are presented graphically in Figure 1.4b. The oscillations become progressively smaller, due to the presence of the exponential term in the solution of equation 1.6,

$$W_o = W[1 - (1/a)e^{\{-Fw_n t\}} \sin(aw_n t + P)]$$
$$P = \tan^{-1}(a/F) \tag{1.10}$$
$$a = [1 - F^2]^{(1/2)}$$

Thus, oscillations in this situation are damped, superimposed over an exponential decay term.

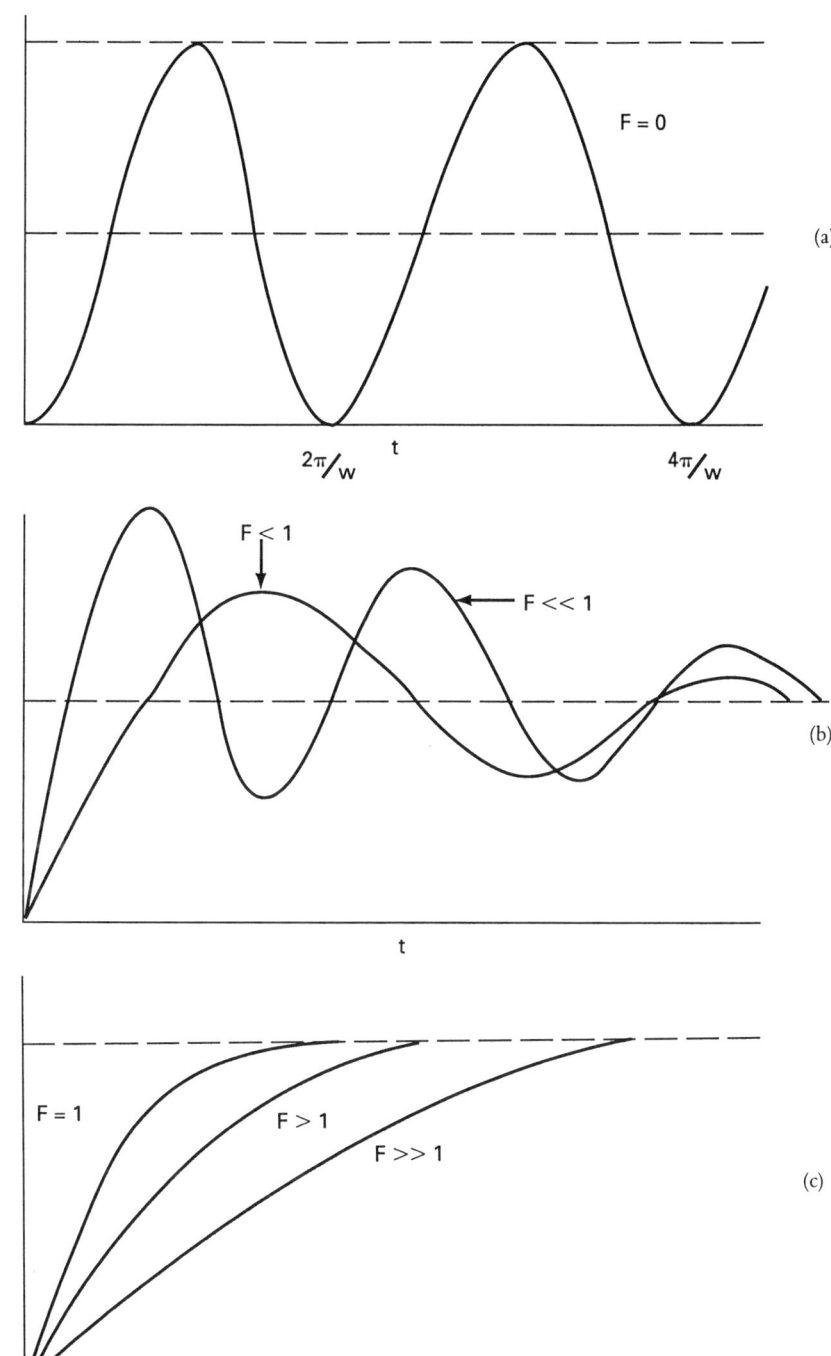

Figure 1.4 Step response of a second-order system. The three plots are the different responses obtained depending on the magnitude of F, whether it equals zero, or greater than zero, or less than zero.

2. If $F > 1$, the two roots are real, so that the solution gives an exponential approach to the desired value. This is depicted in Figure 1.4c. The approach becomes progressively slower, as the damping factor is increased. For a second-order system the gain is such that the response is a damped sinusoid. With an increase in gain, a large number of oscillations appear in the response.

3. When F equals 1, the roots are real and equal. The solution of the homogeneous equation is

$$W_o = e^{-p_i t}\{c + dt\} \tag{1.11}$$

to which a particular solution W_n, should be added for the general solution. c and d are constant parameters of the system. In this case, Equation 1.7 no longer represents the most general solution. The solution, when F equals unity, denotes the fastest response of the system that does not exhibit any oscillation. This condition is known as *critical damping*.

4. Finally, when F equals zero, the roots of characteristic equation are imaginary. The complimentary solution of the differential equation for imaginary roots becomes

$$W_o = C \cos w_n t + D \sin w_n t \tag{1.12}$$

STEP INPUT RESPONSE OF A THIRD-ORDER SYSTEM _____

For a third-order system, additional time is required to transmit changes in the input through the system, and this can result in a dramatic change in the behavior. The differential equation for the step input response of a third-order system can be written as:

$$\{d^3 R_o/dt^3\} + A\{d^2 R_o/dt^2\} + B\{dR_o/dt\} + CR_o = CR_i \tag{1.13}$$

The solution of Equation 1.13 can be expressed in a manner similar to the second-order system as:

$$R_o = R_i + Ae^{\langle p_1 t\rangle} + Be^{\langle p_2 t\rangle} + Ce^{\langle p_3 t\rangle} \tag{1.14}$$

where p_i with $i = 1, 2, 3$ are the eigenvalues of the characteristic equation,

$$p^3 + kp^2 + mp + n = 0 \tag{1.15}$$

k, m, and n are constants, characteristic of the system, related to the system parameters.

As before, the various possible responses given by Equation 1.15 depend on the sign and magnitudes of the eigenvalues. Since this characteristic equation is cubic, one of the roots must be real, while the other two can both be real, or both imaginary, or both complex. Therefore, there are three possible forms of step input response, depicted by Equations 1.14.

1. If all roots are real,

$$R_o = R_i\{1 + Ae^{gt} + Be^{ht} + Ce^{it}\}$$

2. If one root is real, and the other two are imaginary,

$$R_o = R_i\{1 + Ae^{gt} + B\sin(wt + C)\}$$

3. If one root is real, and the other two are complex,

$$R_o = R_i\{1 + Ae^{gt} + Be^{ht}e^{\langle\sin(wt+C)\rangle}\} \tag{1.16}$$

In all Equations 1.16, A, B, and C are constants, whose values again depend on the system parameters. g, h, i, and w are related to the eigenvalues. When the roots are real and negative, the first of Equation 1.16 contains an exponential decaying term and also an oscillation term with constant amplitude. This is shown in part B of Figure 1.5. The last of Equation 1.16 contains a damped oscillation term, superimposed over an exponential decay term. This kind of response is depicted in Figure 1.5C.

Any disturbance to an unstable system will initiate oscillations that increase with time. In the unstable case, with the amplitude of oscillations becoming larger with time, there should be a point at which the amplitude ceases to increase any further. Either the system saturates at this point and thus limits the amplitude of oscillation, or a disruption occurs and components of the system break down. The exact ultimate fate of the system is not of immediate interest. However, one should know the aspects of the system which tend to promote such instability. Instability in physiological systems does not always lead to periodic behavior. They are seldom sinusoidal.

The mathematical models of physiological systems are usually studied with the aid of a computer, since equations to be solved for these models are often complicated and it is difficult to obtain analytical solutions. On an analog computer, the electric circuit is specified directly by the block diagram, and thus corresponds to the real system being modeled.

The study of nonstationary biological processes is accomplished by preliminary signal processing of the signals or the measured variables. Some of these preliminary signal processing methods are (1) signal averaging, (2) Wiener filtering, and (3) selective averaging (Basar, 1976). An examination of the oscillations shows how the oscillations of several variables are related to each other. Two simple propositions are usually used. (1) The component processes are connected so that the output of one serves as the input for the next, and this relation persists around the feedback loop. (2) It takes time for a signal to propagate through a process, so that the signal will appear delayed in phase compared to the input.

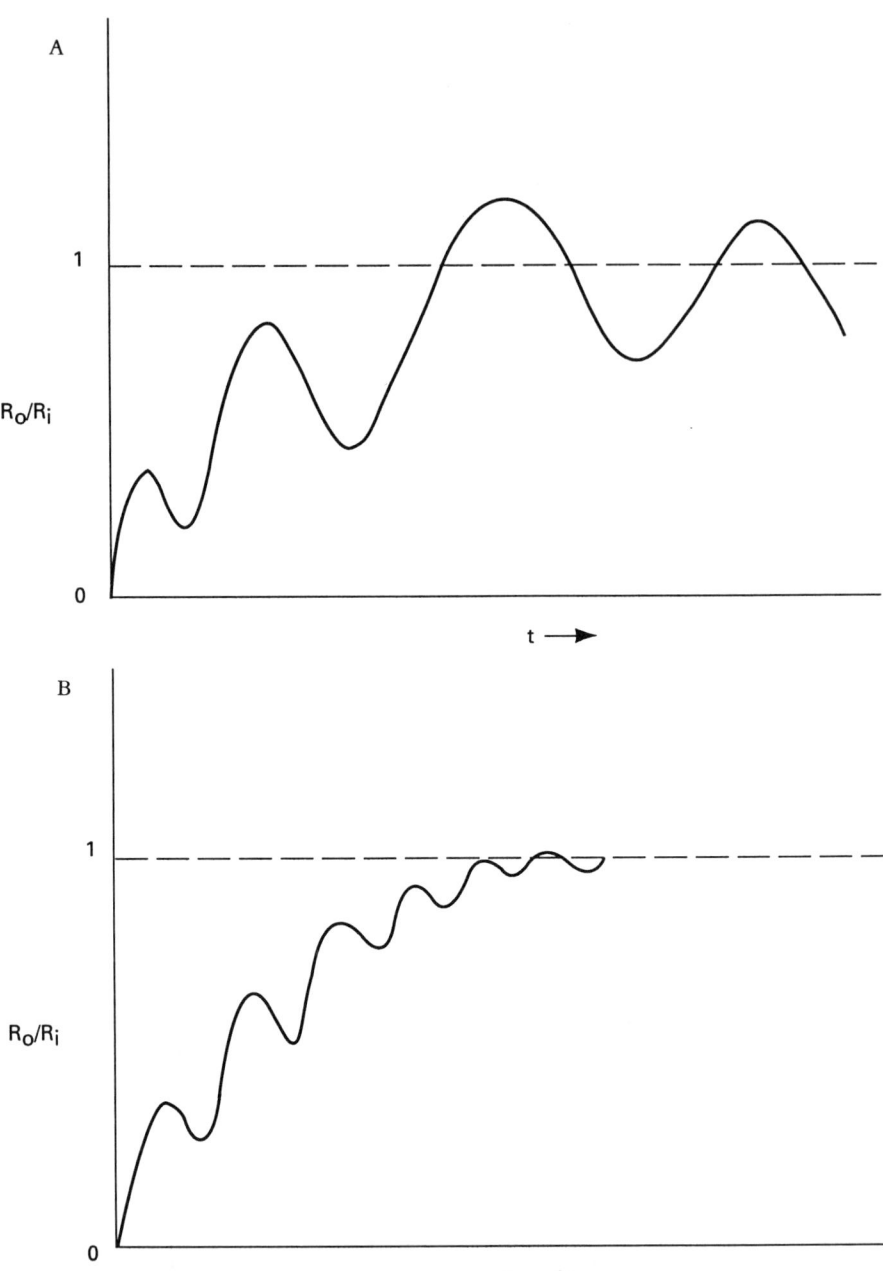

Figure 1.5 Step response of a third-order system. A) represents the step response of a third-order system, whose roots of the characteristic equation are purely imaginary. A continuous oscillation is superposed on the exponential approach to the desired value. B) indicates a damped oscillatory form of a step response of a third-order system. C) shows an unstable form of step response of a third-order system.

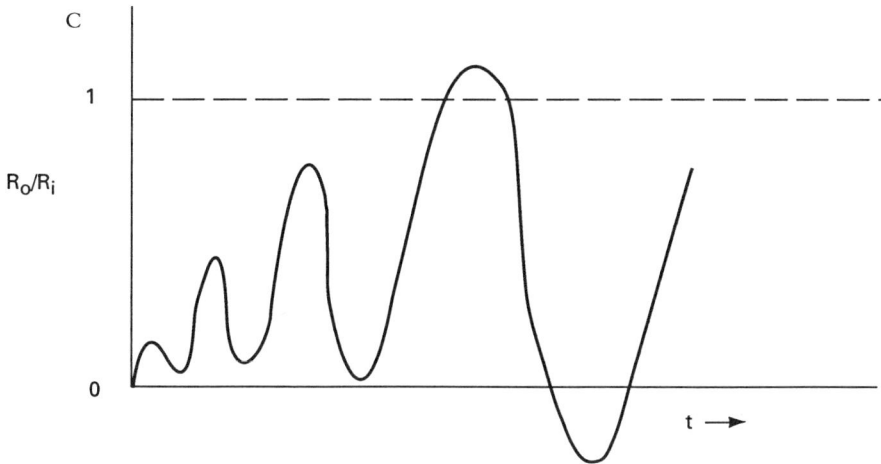

Figure 1.5—(*Continued*).

Any wave form can be represented to an arbitrary degree of accuracy by a sum of sine waves. In addition linear systems obey the principle of superposition. Therefore, the response of a linear system to an arbitrary wave-form signal can be deduced if the system response is known to sine-wave signals of all frequencies. Thus, for a stimulus of a sine wave to the system,

$$x(t) = A \sin t$$

one has the response, $y(t)$ in the frequency domain as

$$Y(s) = H(s)[A\{s/\langle s^2 + w^2 \rangle\}]$$

where $[s/\langle s^2 + w^2 \rangle]$ is the Laplace transform of $\sin t$. The Laplace transform is a mathematical device for changing the variable time t in equations of the system to a new variable s. Operations of differential equations and integration of such equations which are difficult and laborious are converted into algebraic relations. The usefulness of the Laplace transform is based on the fact that no function has more than one transform and no transform corresponds to more than one function.

In addition, Laplace transforms obey the normal algebraic rules, so that the equations can be manipulated algebraically in such a manner that on being transformed back to time function, the solution of the differential equation in question is obtained. If a system is described by a first-order differential equation of the kind

$$S\{dy/dt\} + y = k_1 X_1 + k_2 X_2 + \ldots \tag{1.17}$$

where y is the output and X_1, X_2, . . . and so on, are the inputs corresponding to a block diagram, the solution may be obtained by the technique of the Laplace transform.

A time-domain variable $f(t)$ can be expressed as a new variable $F(s)$ in the complex frequency domain s defined by the Laplace transform relation

$$L[f(t)] = F(s) = \int_o^\infty f(t)e^{\langle -st \rangle} \, dt \qquad (1.18)$$

where $L[\ ]$ means the Laplace transform of $[\ ]$. Certain restrictions exist on the forms of $f(t)$ and there exist reverse relations of transforming $F(s)$ to $f(t)$. Tables of Laplace transforms are available in many books.

The transfer function $G(s)$ is defined as the ratio of the Laplace transform of the input function over the Laplace transform of the output function. If the system has a transfer function given by the expression

$$\langle W_o(s) / W_i(s) \rangle = G(s) = [A(s)/B(s)] \qquad (1.19)$$

where $A(s)$ and $B(s)$ are polynomial functions of s, the step function response of the system will be obtained by taking the inverse transform of the expression

$$W_o(s) = \{[W_i \cdot A(s)]/[s \cdot B(s)]\} \qquad (1.20)$$

where W_i is the magnitude of the input step. When the variables depicted by the input and output pathways are represented in the time domain, for a linear system, one obtains the output (1) by taking the Laplace transform of the input, (2) multiplying it by the transfer function of the block, and then (3) taking the inverse transform of the product to obtain the output.

Experimental studies of biological systems have revealed certain methodological points and general concepts. In order to interpret the results, use has been made of systems theory. The main advantage of the use of systems theory in biology appears to be that when there are different components of the system which contribute to the systems' response, the observer cannot distinguish these components without further analysis. In addition, the presence of a large number of peaks in the transient response do not necessarily reveal the existence of a large number of components.

A linear system can be fully described either in the time domain or in the frequency domain, with its characteristics. All information concerning the frequency characteristics of a linear system is contained in the transient response of the system. All information concerning the time response of the system is likewise contained in the frequency characteristics of the system. Therefore, the stimulus-response relationships can be expressed either in the time domain or the frequency domain. The frequency-domain description has the advantage both in terms of analytical convenience and in terms of often enhancing the information about the signals in the presence of noise. Examples are many in physiological systems where the information of the signals is dependent on their frequency compositions. The description of signals in the frequency domain is important for acoustic and visual signals, since noise can be conveniently filtered out by working in the frequency domain, when the frequency spectrum of the noise does not overlap with the spectrum of the signal.

Feedback may bring about modifications in the system behavior. In a first-order system, defined shortly, the feedback serves to reduce the effective time constant. In general, the effect of the feedback is to increase the speed of response of the system to imposed disturbances. The method of studying frequency characteristics consists of the measurement of the amplitude and phase characteristics of the system under investigation. For these measurements, one applies a sinusoidal signal, having a frequency f to the system. After certain time sufficient for the dampening of the transient, only forced oscillations will remain having the frequency of the signal. The frequency, amplitude of the signal, the amplitude of the output, and the phase difference between the input and the output are measured. The frequency of the input signal is gradually increased to the frequency where the amplitude of the output becomes so small that the instrument cannot record any further. The plots of the ratio of the output amplitude to the amplitude of the input signal represent the amplitude and phase characteristics of the system. The main advantage of the frequency-domain analysis is the fact that the integral equations relating stimulus to response in the time domain become algebraic equations in the frequency domain. This vastly facilitates the descriptions of composite systems, yielding an easy way to combine subsystems.

QUALITATIVE REMARKS ON STABILITY

The question of stability or instability becomes an important subject in systems with feedback, since all such systems may become unstable in the absence of feedback. The homeostatic processes of biology are normally stable, as is evidenced by their continued existence, having acquired stability under evolutionary pressures. The mere existence of stability in a physiological system leads to questions of how that stability is attained and maintained, since the potential for instability always exists.

The basic concept of stability is associated with the notion that a bounded (finite) disturbance to a stable system will produce a bounded response. If the system is stable, then the response will remain small for all small disturbances. On the other hand, any small disturbance to an unstable system will initiate an unbounded response. The unstable system exhibits an irreversible change in its behavior following a disturbance, and in many cases this new behavior causes the system variables to assume values beyond their physical capability.

A closed-loop system becomes an open-loop system if a break is made in the information flow within the loop. Such operations facilitating simplification of the study of the subsystems cannot be easily accomplished in biological systems. Closed-loop control can usually provide compensation against system disturbances. If the controller is properly designed, the dynamic performance can be improved. For a given system, the loop-gain will be an important factor in connection with instability. At a critical value of the gain, a transient once initiated perpetuates itself and oscillations persist. The unstable system is distinguished by the fact that it will continue to oscillate even after an initial disturbance has disappeared, since the signal can grow in magnitude and continue to propagate around a closed loop. A basic upper limit of gain always exists in any particular situation. Thus, a closed loop can eventually become unstable and break into sustained or even destructive oscillations. Thus, in any controlled system, a trade-off between stability and speed of response is usually effected. Manipulation of loop gain is not always possible in a biological system.

Instability is dependent on the properties of the system and is not normally a function of the specific disturbance. If the system is stable, all transients will ultimately disappear, regardless of the disturbance causing them. Any disturbance to an unstable system will initiate oscillations that increase in amplitude with time. All feedback systems are continually beset by disturbances having a large variety of temporal forms. Instability in physiological systems does not always lead to a periodic behavior, but oscillations are not uncommon. Motivation for concern over stability is quite different in engineering systems from that generated by a study of organisms. In engineering design, stability is a prime concern. In most cases, components may be added to the basic system to insure stability. In contrast, homeostatic processes are normally stable except in pathological situations.

Potential instability has immediate and practical consequences for the physiological systems. The boundary between stability and instability in living systems as well as the results of stability analysis of mathematical models is of significant interest. With a given definition of stability, one can develop the criteria for the existence of stability, expressed as mathematical relations between system parameters of the chosen model. In the face of potential instability, what criteria can serve to predict it? What aspects of the system tend to promote instability? Are means available for preventing instability and promoting stable operation over all possible operating regions? An increased gain reduces the sensitivity of the system. However, the penalty that may be associated with the higher gain is the potentiality of the system becoming unstable.

A system is said to be *asymptotically stable* if, following a disturbance, the system returns asymptotically to its previous steady state or, if the disturbance is prolonged, to a new stable stationary state. An *absolutely stable* system is one that remains stable for all values of its parameters. A system is said to be *conditionally stable* if it is stable for intermediate values of gain. The term *orbital stability* refers to the ability of an oscillating or periodic system to maintain its frequency and amplitude. The term *global stability* refers to the ability of a system to remain stable for all possible normal operating conditions. The preceding definitions are taken from mathematical definitions of stability of solutions of certain nonlinear differential equations, presented in Chapter 6.

All feedback systems are continually beset by disturbances having a large variety of temporal forms, and any signal wave form can be resolved into a frequency spectrum by means of a Fourier transform. The Fourier spectrum of an arbitrary time signal is continuous and contains all the frequencies. The conditions under which an active control system can go into an unending oscillating state is of immense interest. It is clear that this oscillation results because of the phasing of input–output relations of the loop elements such that the amplifier pumps energy into the system during each cycle, similar to a child working up oscillations in the swing. When the input of additional energy into the system is discontinued, the swing continues to oscillate, except that the amplitude of the oscillation may gradually decay in time. The swing is an example of an almost conservative system. The system is passive when the child sits on it quietly, but it is actively controlled when the child manipulates the swing. A *conservative system* is one in which the total energy remains constant, without external supply. In an ideal conservative system, the sustained oscillations are essentially sinusoidal, both in angle and in conversion of energy forms from potential to kinetic and vice versa. Most biological systems show an intrinsic oscillatory behavior. There are a large number of biological systems where the oscillatory character plays a major role in their functional nature. The subject of biological rhythms is presented in Chapter 9. A survey of the essentials of the mathematics of oscillations is presented in Chapter 6.

One may view the circulation system as a communication channel containing thermal and hormonal signals. Blood flow is the principal means by which heat is transferred throughout the body. This is a thermal signal. Hormonal concentration is a chemical signal, and the finite time for circulation from the source to the target organs results in a transportation lag. Diffusion may alter the signal profile (Beltrami, 1987) with the result that the signal is transmitted with some distortion. In the absence of a transport lag, a first-order system can never become unstable. A feedback system containing only transport lag may lead to the stability of an otherwise unstable system. In the case of such disorders as glaucoma, hypertension, diabetes, and hypothyroidism, oscillations are normally observed but the system variables leave their normal operating regions and assume abnormal values. The observation of a final state, either one of oscillations or of saturation, does not itself reveal how the system reached that state. One can infer only that instability is possible, but not the cause.

In man-made control systems, the differentiating and integrating devices or parameters of the system are adjusted so that the performance of the system meets certain specifications while working under prescribed conditions. If these conditions change, the performance may become inadequate. If the system is made to be self-adaptive, the parameters are made to change automatically in such a manner as to restore the performance to the best desirable value possible.

When there are many inputs and many outputs, it is possible that all inputs fluctuate, and on some of them it is simply not possible to have any control. Thus, to obtain some kind of an optimum output, it is necessary that the system controls whatever inputs it can, so that these may affect the uncontrollable input values. Part of the nervous system acts as a controller in the regulation of the activities of the organs concerned with the circulation of blood and the digestion of food. These kinds of self-adaptive systems are nonlinear in that the parameters also vary with time.

Evidently, control requires a means of assessing the performance of the system with respect to some optimum value, and a means of comparing the difference between the actual output and the optimum output, or the output of a previous time. This implies the existence of a memory device. Finally, the system should be able to change its parameters or the values of controllable inputs in a direction which will result in values closer to the optimum desired value. Living systems receive inputs from several sources and are self-adaptive.

Nature with scant regard for the comfort and convenience of a theoretician has devised most problems in biology with fully nonlinear properties. The problem of the stability of the solutions of these nonlinear problems presents difficulties even to the present day. The tools available in the past were those that belonged to the linear algorithm. In recent years, a better understanding of the nonlinear phenomena and properties of nonlinear differential equations has been achieved. The approximation methods of either neglecting or mitigating the influence of nonlinear terms have been demonstrated to be of little use. Fortunately, there are examples of fully nonlinear problems in which the solutions can in some sense be written as a "superposition" of modes. Certain nonlinear problems have a simple underlying structure and can be solved by essentially linear methods. These forms are typically in the form of evolution equations. Chaotic behavior should be observable in systems whose dynamics are dominated by relatively few spatial modes. If the system had the property of sensitivity to initial conditions, then the dynamics would be such that the orbits through neighboring points would diverge from each other at an exponential rate.

REFERENCES

ADEY, W.R., and A.F. LAWRENCE, *Nonlinear Electrodynamics in Biological Systems*. New York: Plenum Press, 1983.

BASAR, E., *Biophysical and Physiological Systems Analysis*. Reading, MA: Addison-Wesley Publishing Co., 1976.

BAYLISS, L.E., *Living Control Systems*. San Francisco: W.H. Freeman and Company, 1966.

BELTRAMI, E., *Mathematics for Dynamic Modeling*. New York: Academic Press, 1987.

CAIRNS-SMITH, A.G., *The Life Puzzle*. Edinburgh: Oliver and Boyd, 1971.

COLINVAUX, P., *Why Big Fierce Animals Are Rare?* Princeton, NJ: Princeton Science Library, 1978.

FROHLICH, H., *Biological Coherence and Response to External Stimuli*. Berlin: Springer-Verlag, 1988.

GOLDMAN, S., in *Mineral Metabolism*, vol. 1, part A, p. 61, eds. Comar and Bronner. New York: Academic Press, 1961.

GRODINS, S.F., *Control Theory and Biological Systems*. New York: Columbia University Press, 1963.

JONES, R.W., *Principles of Biological Regulation*. New York: Academic Press, 1973.

KACSER, H., and J.A. BURNS, *Symp. Soc. Exptl. Biology*, vol. 27, (1973), p. 65.

KAISER, F., in *Biological Coherence and Response to External Stimuli*, ed. H. Frohlich. Berlin: Springer-Verlag, 1988.

KALMUS, H., *Regulation and Control in Living Systems*. New York: John Wiley & Sons, 1966.

MARMARELIS, P. Z., and V.Z. MARMARELIS, *Analysis of Physiological Systems: The White Noise Approach*. New York: Plenum Press, 1978.

MILSUM, J.H., *Biological Control System Analysis*. New York: McGraw-Hill Book Co., 1966.

NICOLIS, G., and I. PRIGOGINE, *Self-Organization in Nonequilibrium Systems*. New York: John Wiley & Sons, 1977.

REINER, J.M., *The Organism as an Adaptive Control System*. N.J.: Prentice Hall, 1968.

RIGGS, D.S., *Control Theory and Physiological Feedback Mechanisms*. Baltimore: The Williams and Wilkins Co., 1970.

VAIDHYANATHAN, V.S., in *Molecular and Biological Physics of Living Systems*, p. 147, ed. R.K. Mishra. London: Kluwer Academic Publishers, 1990.

chapter 2

Thermodynamics

Certain laws of nature including those of thermodynamics impose certain constraints on the behavior of biological systems, playing a role in the regulation and control of the system's behavior. The intake of energy in the form of food and its release in metabolic activity are essential for living systems. Many of the biochemical reactions proceed if and only if catalyzed by enzymes. Other reactions cannot proceed spontaneously unless coupled with other spontaneous chemical reactions that can provide the necessary free energy for the synthetic process to proceed. Living organisms transfer chemical energy into mechanical energy, and thermodynamic relations are vital for the control of such processes. An account of the kind of impositions placed by the laws of thermodynamics on the systems is presented in this chapter. In such a presentation, it is presumed that the reader has prior familiarity with the subject of thermodynamics.

Thermodynamic descriptions may be classified as an overall macroscopic approach. In classical mechanics, the transition from statics to dynamics is effectuated by an appeal to Newton's laws of motion. In thermodynamics, the initial development was restricted to systems in states of equilibrium, when the variables and functions of thermodynamics are essentially independent of time. All reversible processes permit adequate description, while irreversible processes, if they are not complex, may be sometimes amenable to adequate description. A detailed discussion of a physical process deals with quantitative statements about every step in the process. In an overall discussion, one is content to restrict all quantitative assertions to the process as a whole.

Physicochemical systems are described in terms of a few simple measurements capable of being performed by means of macroscopic (large-scale) instruments. A description of this kind will be considered adequate if it is such that whenever its numerical data reoccur, the whole measurable subsequent course of the system is reproduced, provided the environment of the system is the same as before. However, usually our inability to find an adequate description is due to the complexity of the process under consideration.

There are four laws of thermodynamics. The zeroth law defines the temperature, T, by appealing to thermal equilibrium. The concepts of free energies, as distinct from internal energy, arose because all the energy available in a system cannot be extracted to do useful work. Only part of the internal energy is available for utilization. The unavailable part of this energy is found to be proportional to temperature, and this results in the recognition of the concept of entropy.

---

† According to Gibbsian thermostatics, one has for reversible processes,

$$dS = \{E/T\} + \{p\,dV/T\}$$

This equation contains no specific reference to properties of the substance. It has the sense of a universal character.

$$\{1/T\} = f_1(E, V)$$
$$\{p/T\} = f_2(E, V)$$

With the use of a thermometer and a pressure gauge, one may establish the validity of the preceding empirical relations. Hence, the equation of state is of the form:

$$V = V(p, T)$$
$$E = E(p, T)$$

The existence of entropy as a state function requires only that the integrability condition

$$\{\partial(1/T)/\partial V\} = \{\partial(p/T)/\partial E\}$$

be satisfied. One has, therefore,

$$S = S(E, V)$$
$$\{\partial S/\partial E\} = (1/T); \quad \{\partial S/\partial V\} = (p/T)$$

The first of the preceding equations can be represented as a surface in a space spanned by the variables S, E, and V. This space is called the Gibbsian space. Thus, two state functions associated with equilibrium exist, namely, E and S. Gibbs postulated the extremum principle: *In an isolated system, the energy tends to a maximum at constant entropy.* A paradox, however, exists as pointed out by Tisza (1966). If an isolated system is not in equilibrium, one can associate no entropy and if it is at equilibrium its entropy can no longer increase.

The principle of conservation of energy is the essence of the first law, while the definition of entropy, S, is accomplished by the second law of thermodynamics. Unlike temperature, pressure, or internal energy, the quantity of entropy is not directly measurable. The change in entropy is calculated using the third law of thermodynamics. To this end, the third law of thermodynamics is stated as *the entropy of a perfectly crystalline substance at absolute zero degree temperature of Kelvin scale, is zero.* One computes the entropy change by the integration of the ratio of specific heat and temperature over the temperature range. The second law of thermodynamics implies that entropy and free energy are not conserved properties of the system. The internal energy, E, and the Helmholtz free energy, F, are related to each other by the relation, $F = E - TS$. Since experiments are easily done under the conditions of constant pressure rather than constant volume, the thermodynamic functions enthalpy, H, $(H = E + pV)$ and the Gibbs free energy, $G = F + pV$ are defined.

Thermodynamics is a phenomenological theory concerned exclusively with the macroscopic properties of physical systems. It deals with certain properties of the system and their relationships to the variables of the system. The main importance of thermodynamics in biology is that it can provide information as to how a given biological system may be functioning. Since the formulation of the second law of thermodynamics, attempts have been made to enlarge the realm of classical thermodynamics to include irreversible processes.

Classical thermodynamics is limited to considerations of systems in equilibrium. For an irreversible process occurring in a closed system, the second law requires that the change in entropy of the system is greater than the heat absorbed by the system divided by the absolute temperature of the surroundings. The second law for a closed system may be written that the entropy change is the sum of two terms: (1) the change in entropy due to heat absorbed from the surroundings, and (2) the change due to entropy production within the system. The second law requires that the second term be positive definite. In an open system, the first term should contain the changes in entropy resulting from matter being transported in and out of the system. In this situation, it is possible for the first term as well as the total entropy change to be either positive or negative.

The function of entropy is a pivotal concept in both equilibrium and nonequilibrium thermodynamics. However, in extending the concept of entropy to nonequilibrium states, one has to take into account the fact that the local entropy depends not only on the local values of thermodynamic variables but also on the values of such quantities in adjacent locations, since in the nonequilibrium state, gradients of local parameters exist. The production of entropy and its flow is nonzero in the nonequilibrium state. The rate of entropy production is zero in the equilibrium state, while it is minimum in the nonequilibrium stationary state.

The familiar variables for the description of systems in thermodynamics are the temperature, volume, pressure, and concentrations of the various components of the system for a multicomponent system. These variables may be broadly identified as macroscopic variables and may be further classified as either intensive or extensive variables. Temperature and pressure are macroscopic variables and their significance is meaningful in terms of measurements made on matter in bulk. These terms lose their meaning at the atomistic or molecular level. These also do not have well-defined meanings in an extremely inhomogeneous system not in equilibrium.

Thermodynamics describes physicochemical systems in terms of a few simple measurements capable of being performed by means of macroscopic instruments. Thus, thermodynamic theory is developed in close contact with measurements. Solution thermodynamics may be classified into three major aspects: (1) equilibrium aspects, (2) approach to equilibrium aspects, and (3) the study of transport phenomena. A system is said to be in equilibrium with respect to a given change of state when the free energy of the system, expressed as a function of the extent of that change of state, is a minimum.

The change in free energy of the system produced by any change of state may be expressed by the sum of the products of the chemical potential, or partial molal free energy of unit quantity of each component and the quantity of that substance produced by the change of state. Quantities of the substances that disappear are counted as negative quantities produced. The study of a very large number of reactions is thus reduced to the study of the potentials of a much smaller number of substances. The potentials in real systems are expressed as differences from those in an idealized standard state.

Symbolically, these may be expressed as

$$\sum_i v_i G_i = 0 \qquad (2.1)$$

where $\sum$ stands for the sum over all the z_i reactants and products, $i = 1, 2, \ldots, z$, v_i is the number of units of component i produced in the reactions, and G_i is the potential of component i. The components of the system are the molecular species that make up the system, which may be added to or removed from any one of the phases of the system. A solution is a single phase that contains more than one component. In order to describe the system, one needs to know the volume V of the system, its pressure p, and its temperature T, as well as the quantity (in moles n_i) of each of its components.

Some properties of a system, such as its mass or volume, may be taken as a measure of its quantity. Other properties such as the composition or density when pressure and temperature are maintained constant are independent of the amount. Those properties which are independent of the amount of the system are called the *intensive properties*. Those properties of the system that are proportional to the quantity present in the system are called *extensive properties*. The ratio of two extensive variables, as well as the ratio of two intensive variables, is intensive.

The thermodynamic concepts which permit prediction of the maximum efficiency of a heat engine also permit the computation of the maximum yield in a chemical reaction to equilibrium. Another primary objective of thermodynamics is the establishment of a criterion for the determination of the feasibility or spontaneity of a given transformation. The fundamental principles of thermodynamics can be formulated without any recognition of the existence or properties of molecules. In this respect, it is a phenomenological science.

Although the theory of thermodynamics provides the foundation for the solution of many chemical problems, the answers obtained are generally not definitive. Classical thermodynamics is capable of formulating the necessary conditions for a process to occur, but not sufficient conditions. Thermodynamics cannot make any statement regarding the rate of a chemical reaction. Similarly, with regard to the work obtainable from a chemical reaction, only limiting values can be computed.

Classical thermodynamics also encounters difficulties in treating fluctuation phenomena, such as Brownian motion or certain turbidity problems. Thermodynamics cannot explain fluctuations, and in fact, denies even their existence. A fluctuation into a less probable state of a system can be shown to involve a decrease of entropy, in contradiction to nineteenth century ideas on the steady increase of entropy. *Classical thermodynamics is predicated on the assumption that a definite and reproducible value can always be measured for macroscopic properties of a system.* Therefore, thermodynamics can deal with only systems where the number of molecules is so large that random fluctuations are not observable.

The first law of thermodynamics states that energy can neither be produced, nor consumed, but only transformed or transported. (The first law may be stated as: It is impossible to construct a device that operating in a cycle will produce effects other than the performance of work or by an auxiliary mechanical system.) Thermodynamic functions (or potentials) are (internal) energy, E, enthalpy, $H = E + pV$, the Gibbs free energy, G, the Helmholtz free energy, F, the entropy, S, and the partial molal free energy of component i of a multicomponent system, called the chemical potential, μ_i. (We utilize the notation δ to indicate the variation of a quantity that the symbol δ prefixes, and use the prefix d to denote the differential of that quantity. n_i denotes the number of moles of species i present in the system.)

Functions of state are independent of the manner in which a particular state of the system is attained, and it is for this reason that heat and work are not functions of state. Energy is useful only when it is converted into work. The first law of thermodynamics yields a relation between the energy input, the work done by the system, and the energy lost in the form of heat, but it does not provide information as to how much of the energy added will actually be used to do work. The efficiency of energy transforms† into useful work is given by the second law of thermodynamics. The second law establishes that there is only one possible direction of time for irreversible processes. All irreversible processes can only go forward in time, while reversible processes can go forward or backward.

FUNDAMENTAL EQUATIONS OF HOMOGENEOUS OPEN SYSTEMS

The four fundamental relations for open systems are

$$dE = T\,dS - p\,dV + \sum_i \mu_i\,dn_i \tag{2.2}$$

$$dH = T\,dS + V\,dp + \sum_i \mu_i\,dn_i \tag{2.3}$$

$$dF = -S\,dT - p\,dV + \sum_i \mu_i\,dn_i \tag{2.4}$$

$$dG = -S\,dT + V\,dp + \sum_i \mu_i\,dn_i \tag{2.5}$$

GENERAL CONDITIONS FOR EQUILIBRIUM

It follows from the second law of thermodynamics that a system is in a state of equilibrium if

$$\delta S \leq \delta Q / T \tag{2.6}$$

for all virtual processes to neighboring states. T is temperature.

For closed systems, the first law of thermodynamics is of the form:

$$\delta Q = \delta E + \delta W \tag{2.7}$$

For a closed system, energy changes due to transfer of mass between the system and its surroundings are not considered in the first law. Changes in masses of various components within a closed system due to chemical reactions do not involve work done by the surroundings and thus do not

---

† Carnot's main theorem states that all reversible engines operating between two distinct temperatures have the same efficiency, regardless of the nature of the system constituting the working substance. In addition this efficiency is larger than that of any other nonreversible engine.

appear in the expression for the first law. If the external force is a uniform normal pressure p, one may write

$$\delta Q = \delta E + p \, \delta V \tag{2.8}$$

Equations 2.6 and 2.8 lead to the general criterion of equilibrium for closed systems:

$$\delta E + p \, \delta V - T \, \delta S \geqslant 0 \tag{2.9}$$

Special forms of Equation 2.9 valid with additional restrictions are of interest.

1. The criterion of equilibrium for a closed system at constant energy E and volume V is

$$(\delta S)_{E,V} \leqslant 0 \tag{2.10}$$

2. The criterion for equilibrium for a closed system at constant entropy S and volume V is

$$(\delta E)_{S,V} \geqslant 0 \tag{2.11}$$

3. The criterion for equilibrium for a closed system at constant volume V and with a uniform, constant temperature T is

$$(\delta E)_{T,V} - T(\delta S)_{T,V} \geqslant 0 \tag{2.12}$$

From the expression for Helmholtz free energy, F,

$$F = E - TS$$

$$dF = dE - T \, dS - S \, dT$$

one has

$$(\delta F)_{T,V} = (\delta E)_{T,V} - T(\delta S)_{T,V}$$

Thus, Equation 2.12 may be rewritten in the form

$$(\delta F)_{T,V} \geqslant 0 \tag{2.13}$$

4. The criterion of equilibrium for a closed system characterized by a uniform, constant temperature T and a uniform, constant pressure p is

$$(\delta E)_{T,p} + p(\delta V)_{T,p} - T(\delta S)_{T,p} > 0 \tag{2.14}$$

Equation 2.14 can be expressed in terms of the Gibbs free energy, $G = F + pV$, as

$$(\delta G)_{T,p} = (\delta E)_{T,p} + p(\delta V)_{T,p} - T(\delta S)_{T,p} \geqslant 0 \tag{2.15}$$

Thus, the criterion of equilibrium for a closed system of uniform pressure and temperature can be stated as

$$\{\delta G\}_{T,P} \geq 0 \tag{2.16}$$

The total amount of work which can be obtained from a system in a nonequilibrium state will depend on the manner in which the equilibrium state is attained. The question naturally arises as to which way of reaching equilibrium will produce the maximum possible work. The answer to this question is that *the work done by the system is a maximum when its entropy remains constant,* that is, when the transition to equilibrium is a reversible one.

If a closed system is not in a state of equilibrium, then its macroscopic state will vary with time until the system eventually reaches a total equilibrium state. If every macroscopic state of the system is specified by the distribution of its energy between various subsystems, one can state that successive states through which the system passes correspond to energy distributions of greater probabilities. This increase in probability is given by the expression $\exp(\Delta S/k_B)$, whose exponent is an additive quantity, namely, the entropy of the system. (ΔS) is the change in entropy and k_B is the Boltzmann constant. Therefore, processes occurring in a closed system, which is not in equilibrium, occur in such a manner that the system passes from states of lower entropy to those of higher entropy, until finally the entropy attains its maximum possible value, corresponding to the state of statistical equilibrium.

MATHEMATICAL TOOLS OF THERMODYNAMICS

The subject of thermodynamics deals with functions of several variables. The partial derivative of a function $z(x_1, x_2, \ldots, x_n)$ with respect to the variable x_i is denoted by $\{\partial z/\partial x_i\}|_{x_j \neq x_i}$ where the subscript indicates that the derivative is taken assuming that the quantity in the subscript is maintained constant. The widely known relationships existing among the derivatives of functions of several variables are utilized when considering the differential equations of thermodynamics (Sychev, 1983).

Often, one utilizes the reciprocal relation for the partial derivative,

$$\{\partial y/\partial x\} = [1/\{\partial x/\partial y\}] \tag{2.17}$$

This is the *theorem of inverse quantities.* A proof of this theorem is presented in Appendix 2A.

The chain rule of differentiation is also utilized.

$$\{\partial y/\partial x\}_z = \{\partial y/\partial u\}_z\{\partial u/\partial x\}_z \tag{2.18}$$

The Bernoulli-Euler *theorem of equality of mixed second-order derivatives* is utilized to derive certain important equations of thermodynamics. Euler's theorem describes the property of homogeneous functions. A homogeneous function of degree n obeys the equation,

$$nf(x^r) = \sum_{i=1}^{r} x_i \{\partial f / \partial x_i\}\big|_{x_{i(r-1)}}$$

This theorem states that if the mixed second-order derivatives of a function $z(x, y)$ are continuous at a point $p(x, y)$, then they are equal at this point. For the function $z(x, y)$ the value of its mixed second-order derivative does not depend on the order of differentiation, namely,

$$\{\partial^2 z / \partial x\, \partial y\} = \{\partial^2 z / \partial y\, \partial x\} \tag{2.19}$$

The utilization of Maxwell's relations for the derivation of various additional thermodynamic relations is presented in the appendix of this chapter.

THERMODYNAMIC FUNCTIONS OF MIXTURES

Thermodynamic variables can be classified as either extensive or intensive variables. For every extensive variable, say Y, there corresponds an intensive variable denoted by y_i, defined by the partial derivative,

$$y_i = (\partial y / \partial n_i)\big|_{n_{j \neq i}, T, p} \tag{2.20}$$

Such intensive variables are called the *partial molar quantities* associated with the corresponding extensive variable.

The *chemical potential* is a partial molar quantity associated with the free energy G. The chemical potential of a component i, in a multicomponent mixture, varies with the temperature, pressure, and composition and is also dependent on the strength of any externally applied field, such as an electric, gravitational, or magnetic field. There exists an important identity between an extensive variable and the corresponding partial molar quantities called *Euler's theorem*.

$$Y = \sum_{i}^{n} n_i y_i \tag{2.21}$$

Specific cases of this identity related to the partial molar volume and chemical potentials for an n = component system are

$$V = \sum_{i}^{n} n_i v_i \tag{2.22a}$$

$$G = \sum_{i}^{n} n_i \mu_i \tag{2.22b}$$

If in the Gibbs relation,

$$dG = -S\,dT + V\,dp + \sum_i^n \mu_i\,dn_i$$

the differential of Equation 2.22b is inserted, one obtains the *Gibbs-Duhem relation*,

$$-V\,dp + S\,dT + \sum_i^n n_i\,d\mu_i = 0 \tag{2.23}$$

In the Gibbs-Duhem relation, all differentials refer to intensive quantities. Similarly, the n partial molar quantities of an n-component system are related by the identity

$$\sum_i n_i\,dy_i = 0, \quad \text{at constant } p \text{ and } T$$

$$\sum_i n_i\,d\mu_i = 0, \quad \text{at constant } p \text{ and } T$$

Since under stationary-state conditions, the forces acting on any element of volume of a system should balance, one may generalize the last of the preceding equations for the stationary state.

$$\sum_i^n n_i\,\nabla\mu_i = 0 \tag{2.24}$$

This is known as the Gibbs–Duhem relation valid for the stationary state, which is a *force-balance equation*.

A fundamental property of thermodynamic potentials is that the thermodynamic properties of the system are uniquely defined, if the chemical potentials μ_i's of the components of a multicomponent system are known as functions of the independent variables T, p, and n_is, where n_i is the amount of component i present in the system. For a physicochemical change caused by a chemical reaction, the term affinity, $\mathscr{A}$, is defined by the relation

$$\mathscr{A} = -\sum_i \nu_i\mu_i \tag{2.25}$$

where ν_i is the stoichiometric coefficient of component i. The equilibrium of the chemical reaction occurs when affinity equals zero. For systems open to the surroundings for matter flow, the condition for equilibrium is that the chemical potential of a specified species is independent of both position and time variables.

Interest in accounting for the thermodynamic functions of a multicomponent system in terms of the properties of the pure components has led to the concepts of thermodynamic functions of mixing. The mixing of two pure liquids to form a mixture is evidently an irreversible process, since unmixing cannot be done without an additional expenditure of energy. Thus, an entropy of mixing arises.

For a binary mixture, the change in free energy due to mixing of components 1 and 2 is expressed by the relations

$$\Delta G^M = G(p, T, n_1, n_2) - G_1^o(p, T, n_1) - G_2^o(p, T, n_2) \qquad (2.26)$$

where the superscript M indicates thermodynamic functions of mixing. The superscript o is utilized to indicate the corresponding property of the pure component. Using small scripts for denoting quantities per mole (g thus represents free energy per mole), one has

$$\Delta g^M = g(p, T, x_1) - x_1 g_1^o(p, T) - x_2 g_2^o(p, T) \qquad (2.27)$$

A solution will be described as an *ideal solution* if the Gibbs free energy of mixing takes the simplest form,

$$\Delta g^M = RTx_1 \ln x_1 + RTx_2 \ln x_2 \qquad (2.28)$$

R is the gas constant and x_i denotes the mole fraction of component i in the multicomponent system. When two components with no interactions are mixed to form a mixture, the entropy of such a mixing process is nonzero and results in a finite increase in entropy due to mixing. Since mole fractions are positive definite and less than unity, the left–hand side of Equation 2.28 is necessarily positive definite. Thus, for an ideal solution, the entropy of mixing is given by the expression

$$\Delta s^M = -(\Delta g^M / T) = -R\{x_1 \ln x_1 + x_2 \ln x_2\} > 0 \qquad (2.29)$$

In Equations 2.28 and 2.29, R denotes the gas constant. The term in parentheses of the last of Equation 2.29 denotes the ideal solution. An ideal solution will be formed by mixing of two components, 1 and 2, only under two conditions, namely: 1) the intermolecular interactions are of the hard sphere type, that is, the interaction between two molecules is zero beyond a specified distance from the center of a molecule, and is infinite for distances smaller than the size of the molecules, and the molecules of the two components are of similar size. Otherwise, there will be a volume of mixing, which violates ideal solution properties. 2) the intermolecular pair potentials for like and unlike interactions, 11, 12, and 22 interactions are the same and the molecules are of similar size. For ions distributed in a solution, or on a surface, ideality of mixing will never be realized, since, unlike ion attractive interactions and like ion repulsions cause a nonrandom distribution. Random distribution is essential to result in dependence of the chemical potential on the logarithm of mole fractions.

For an ideal solution, changes in the chemical potential of the components with composition upon mixing can be written as

$$\mu_i^M = RT \ln x_i$$
$$\mu_i = \mu_i^o + RT \ln x_i \qquad (2.30)$$

Therefore, the difference between the thermodynamic functions of mixing for real solutions, experimentally measured, and the values corresponding to an ideal solution, at the same temperature, is called the *excess thermodynamic functions*. The excess Gibbs free energy of the mixing of a binary mixture is given by the relation in Equation 2.31, Δg representing the total free energy change upon mixing per mole,

$$\Delta g^E = g - [x_1 g_1 + x_2 g_2 + R T x_1 \ln x_1 + R T x_2 \ln x_2]$$
$$= \Delta g^M - \{R T x_1 \ln x_1 + R T x_2 \ln x_2\} \tag{2.31}$$

The thermodynamic excess functions will differ from the corresponding thermodynamic functions of mixing only in quantities involving entropy. *It should be emphasized that the validity of Equation 2.30 is dependent on the validity of the expression for the ideal solution entropy of mixing.* Thus, for an ideal solution, namely, a solution in which there are no interparticle interactions over distance between particles and molecules behave as hard sphere particles, the chemical potential of a specified component varies as a function of its mole fraction. For real solutions, this is hardly the situation. For dilute solutions, since solvent is present in large excess, one may replace the mole fraction by concentration terms. Since all solutions approach ideality when solute concentrations become extremely small (the system tending to a one-component solvent system), such replacement of mole fractions by concentrations is justified only for an extremely dilute solution. The concentrations of solute, especially electrolytes, in biological systems are markedly high. Therefore, the solutions of biological systems cannot be regarded as anywhere close to ideality. This point raises serious questions regarding the validity of approximations frequently made both in the theory as well as in the application of such simple theories to the interpretation of experimental data. Behavior of such complex systems cannot be explained on the basis of simplified theories.

The variation of the chemical potential of a specified component, say i, with the variation of concentration of another species, say j, in a multicomponent system at constant pressure and temperature is the fundamental yet unsolved problem of physical chemistry. Since real solution properties differ from the expected behavior of ideal solutions, such deviations are expressed (but not explained) by the introduction of the concepts of *activity* and *activity coefficients* in physical chemistry.

The chemical potential of component i for real solutions is expressed as

$$\mu_i = \mu_i^o(T, p) + R T \ln\{\gamma_i x_i\} \tag{2.32}$$

where γ_i is the activity coefficient of component i. It is related to concentration C_i and activity a_i of component i by the relation

$$a_i = C_i \gamma_i \tag{2.33}$$

The ideal solution expression for the chemical potential of species of kind i, in dilute solutions,

$$\mu_i = \mu_i^o(T, p) + RT \ln C_i \tag{2.34}$$

is evidently approximate, and is not likely to be valid for any real systems. Equation 2.30 is obtained from calculation of the ideal entropy of mixing. Since free energy comprises both entropy and energetic terms, in place of Equation 2.32, one may assume the validity of the expression (Vaidhyanathan, 1979, 1980) for chemical potential of species k in real solution as

$$\mu_k(x) = \mu_k^*(T, p) + Z_k e \phi(x) + kT \ln C_k(x) + \sum_j C_j(x) H_{kj} \tag{2.35}$$

where H_{kj} are molecular integrals over the potential energy of interactions of species k with jth kind of molecules. $C_k(x)$ represents the concentration of species k, expressed in moles per unit volume at location x, in the inhomogeneous region of the system. The last term of Equation 2.35 denotes the correction to the ideal dilute solution value of the chemical potential of species k, when its activity coefficient does not equal unity. We have written the expression such that it is valid for an inhomogeneous region where concentrations can be functions of the position variable x. Also, it is implied that species k can be a charged ionic species, with valence charge number Z_k and charge $Z_k e$, when the electric potential felt by a unit charge placed at location x equals $\phi(x)$. k_B is the Boltzmann constant. The molecular integrals H_{jk} may be regarded as constants, independent of position variable. One may factor the sum over species into two kinds, one over the charged species and the other over the uncharged species. Since interactions between ionic species involve the product of the charges, the molecular integrals denoting the ion–ion interaction terms may be expressed further as $H_{kj} = Z_k Z_j e^2 H$, where H is a charge-independent term. Additional considerations of Equation 2.34 are presented later in Appendix 3.

SYNOPSIS OF CLASSICAL THERMODYNAMICS _____

In terms of local variables, the Gibbs relation for unit mass at a location of the system is

$$T \, dS = dE + p \, dV - \sum_i \mu_i \, dw_i \tag{2.36}$$

where the quantities of entropy S, internal energy E, volume V, and the mass fraction $w_i = (m_i/m)$ are the local quantities per unit mass. μ_i is the chemical potential of the ith component. The differentials are first-order approximations to the changes in the preceding quantities at a given location with the variation of time.

In spite of the universal validity of the second law of thermodynamics, the calculation of entropy changes in specific cases can be elaborate. Sometimes it is convenient to find other functions of state which are related to real measurable quantities. A large number of functions of state can be generated from internal energy by the addition or subtraction of variables. The choices of such variables are dictated by the kinds of constraints imposed on the system. Under conditions of constant temperature and pressure, it is convenient to introduce the Gibbs free energy of the system. Almost all thermodynamic functions can be computed from work content, or the Helmholtz free energy of the system. For example, the pressure p, is

$$p = -\{\partial F/\partial V\}_{T,N} \tag{2.37}$$

where N is the total number of molecules. The entropy S is given by

$$S = -\{\partial F/\partial T\}_{V,N} \tag{2.38}$$

The energy E given by the expression $E = F + TS$ is a function of temperature alone. The chemical potential of a component, k, μ_k is defined by the expression

$$\mu_k = \{\partial F/\partial N_k\}_{T,V,N} \tag{2.39}$$

The heat capacity at constant volume, C_V, and the heat capacity at constant pressure, C_p, are defined by the relations

$$C_V = \{\partial E/\partial T\}_{V,N} \tag{2.40}$$

$$C_p = \{\partial H/\partial T\}_{p,N} \tag{2.41}$$

where H is the enthalpy, defined by the fundamental relation

$$H = E + pV \tag{2.42}$$

The coefficient of thermal expansion α, and the isothermal coefficient of compressibility β are defined by the relations

$$\alpha = \{\partial \ln V/\partial T\}_{p,N} \tag{2.43}$$

$$\beta = \{\partial \ln V/\partial p\}_{T,N} \tag{2.44}$$

NONEQUILIBRIUM SITUATIONS

When equilibrium does not prevail in a system, as is the case in living systems, it becomes essential to define precisely the concepts of temperature, pressure, and entropy, since these are macroscopically averaged quantities, and these do not have well-defined meanings at the molecular level. In order to extend the realm of the meanings of quantities of equilibrium thermodynamics to nonequilibrium situations a priori, one has to assume that such quantities indeed exist and have well-defined meanings at least over a small range of volume elements of the system. In this manner, one operates as if the equilibrium condition in each element of volume exists in the nonequilibrium state. This is known as the *assumption of local equilibrium*.

Analytically, the assumption of local equilibrium implies that the Gibbs fundamental relation, expressing the combination of the first and second laws of thermodynamics, is valid in the nonequilibrium state. In addition, it is asserted that the entropy production is positive definite. (This latter assumption is a consequence of Boltzmann's *H*-theorem.) The assumption that the usual thermodynamic functions can be defined for nonequilibrium states is of a more complicated nature, and the range of validity of this assumption can only be determined by consideration of a definite mechanical model of the irreversible process. When an irreversible process occurs in the system, the entropy change will always be positive and nonzero. Thus, the property entropy behaves like a nonconserved fluid.

Three basic postulates are involved in the development and extension of thermodynamics to systems not in equilibrium.

1. The entropy balance equation stated in the following is computed from the Gibbs relation of the equilibrium thermodynamics.
2. The rate of change of entropy production, expressed as the sum of products of existent forces and fluxes, is nonnegative.
3. Each flux can be expressed as a linear combination of forces and the phenomenological coefficients in linear nonequilibrium thermodynamics.

ENTROPY BALANCE AND THE SECOND LAW OF THERMODYNAMICS

Equilibrium thermodynamics describes main features of entropy. The entropy of the system, which is an extensive quantity, is additive. That is, the entropy of the sum of two systems is the sum of their entropies. The first law of thermodynamics stipulates that the energy of an isolated system is constant in time. The second law of equilibrium thermodynamics suggests that the entropy of the system increases as long as changes occur within the system. One may visualize entropy as a fluid, which may be either created or destroyed within the confines prescribed for the system.

The generalized relation for entropy production for a system in which the flows of energy and matter are accompanied by a flow of electricity and the occurrence of a chemical reaction can be written as

$$T\, dS = dE + p\, dV - \sum_i \mu_i\, dw_i - \phi\, de + (1/\rho)\, \mathcal{A}\, d\xi \qquad (2.45a)$$

$$(\partial\rho e / \partial t) + \nabla \cdot J_e = 0 \qquad (2.45b)$$

where ϕ is the electric potential, e is the specific charge (per unit mass), ξ is the extent of the chemical reaction per unit volume, and $\mathcal{A}$ is the affinity of the chemical reaction. The charge conservation relation is expressed by Equation 2.45b.

Therefore, a differential change in entropy can be written as

$$dS = d_e S + d_i S \qquad (2.46)$$

$d_e S$ arises from the exchange of entropy with the system's surroundings and $d_i S$ comes from the internal production or destruction of entropy. The factoring of the total entropy into two parts, $d_i S$ and $d_e S$, can be justified for both closed systems and for open systems, in which matter can be exchanged across the boundary through permeable walls.

Second, $d_e S$ can be positive, negative, or zero, depending on the system's interaction with its surroundings. On the other hand, $d_i S$ is always positive and may equal zero. Therefore, for an isolated system, dS equals zero, since there is no interaction with its surroundings. Therefore, dS is always positive definite for an isolated system, which is the statement of the second law of thermodynamics.

Third, if the system is free to receive heat from a reservoir at temperature T but is otherwise isolated, then

$$dS = (dQ / T) \qquad (2.47a)$$

Thus,

$$dS \geq (dQ / dT) \qquad (2.47b)$$

which is an alternate form of expression of the second law, for isothermal boundary conditions. The *local-balance equation* for entropy is

$$\{\partial(\rho S) / \partial t\} + \nabla \cdot j_S = \sigma_S \qquad (2.48)$$

where ρ is mass per unit volume, and t is time. By virtue of the local-equilibrium conditions, the fundamental relation of equilibrium thermodynamics,

$$T\, dS = dE + p\, dV - \sum_k \mu_k\, dC_k \qquad (2.49)$$

is valid for a system not in equilibrium, for a small volume element in which the thermodynamic variables and functions can be assumed to be constant. It should be recognized that in a system in local equilibrium, processes between neighboring cells are thermodynamically slow. The Gibbs relation must be formulated in the center of the mass frame for the system in macroscopic motion, since equilibrium thermodynamics cannot deal with convective phenomena.

The *dissipation function* is defined by

$$\sigma = T\{dS_i/dt\} = \sum_k J_k x_k \tag{2.50}$$

The calculation of the dissipation function by itself is of little use. Ambiguity exists in the specification of the dimension of the dissipation function. Since entropy is an extensive quantity, the dissipation function defined as the rate of entropy production is asserted to be extensive. However, for matter transport, the flux has the dimension of quantity (mole) transported per unit area per unit time. The corresponding thermodynamic force, namely, the gradient of the chemical potential, has the dimension of energy per mole per unit length. Thus, the product force × flux, $J \cdot X$, yields the dimension for the dissipation function as the energy per unit volume per unit time. This is an intensive quantity, while entropy production per unit time is an extensive quantity. From this it appears that the dissipation function should have the dimension of the rate of entropy production per unit volume.

STABILITY

The equilibrium state of a system represents a stable situation. Since life and the living state as well as homeostatic conditions in biology denote a relatively stable or a quasi-stable situation, the stability concept and analysis are intimately related to the preservation of stationary-state situations in living things.

The conditions for thermal equilibrium in classical thermodynamics are obtained from the conditions of maximum entropy. Requiring that the derivatives with respect to energy and volume equal zero, one obtains the result that the temperature and pressure of a body should be the same in all parts of the body. However, the vanishing of the first derivative is only a necessary condition for an extremum to occur. This extremum can be either maximum or minimum. In order that the extremum represents a maximum, the second derivative should be negative. For a minimum, as in calculus, the second differential evaluated at the extremum point should be positive. Any small perturbation applied to the system, when its energy state is at a minimum, will be nullified and the system returns to the state of stability.

The condition for equilibrium is that the quantity $\{E - T_o S + p_o V\}$ is a minimum. For equilibrium,

$$\{dE - T\,dS + p\,dV\} = 0 \tag{2.51}$$

where E, S, and V are the energy, entropy, and volume of a system and T_o and p_o are the temperature and pressure of the external environment. Obviously, T_o and p_o are the temperature and pressure of the system under consideration for the equilibrium state.

For every small deviation from equilibrium, the quantity, $\{E - T_o S + p_o V\}$ must be positive.

$$\delta E - T_o\,\delta S + p_o\,\delta V > 0 \tag{2.52}$$

Hence, the minimum work which should be done to transform the system from its equilibrium state to a neighboring state is positive definite.

Expanding E in a Taylor series, while considering E as a function of S and V, and retaining only the leading three terms, one has

$$\begin{aligned}
\delta E = {}& \{\partial E/\partial S\}\,\delta S + \{\partial E/\partial V\}\,\delta V \\
& + (1/2)[(\partial^2 E/\partial S^2)\,\delta S^2 + 2(\partial^2 E/\partial S\,\partial V)\,\delta S\,\delta V \\
& + (\partial^2 E/\partial V^2)\,\delta V^2]
\end{aligned} \tag{2.53}$$

Since, $(\partial E/\partial S)V = T$, $-(\partial E/\partial V)_S = p$, the first-order terms are equal to $T\,\delta S - p\,\delta V$, and thus cancel. It follows, therefore, that the first nonvanishing terms should be positive definite. That is,

$$\{\partial^2 E/\partial S^2\}\,\delta S^2 + 2\{\partial^2 E/\partial E\,\partial V\}\,\delta SV + \{\partial^2 E/\partial V^2\}\,\delta V^2 > 0 \tag{2.54}$$

In order that this inequality holds, two conditions should be satisfied.

$$\{\partial^2 E/\partial S^2\} > 0 \tag{2.55a}$$

$$\{\partial^2 E/\partial S^2\}\{\partial^2 E/\partial V^2\} - [\{\partial^2 E/\partial S\,\partial V\}]^2 > 0 \tag{2.55b}$$

$$\{\partial^2 E/\partial S^2\} = (\partial T/\partial S)V = T/C_V \tag{2.55c}$$

Since, $\{T/C\}$ is positive and T is positive, it follows that the specific heat at constant volume, CV, should be positive. This is equivalent to the condition that

$$\{\partial p/\partial V\}_T < 0 \tag{2.56}$$

Therefore, an increase in pressure at constant temperature always results in a decrease in volume. The inequalities expressed in Equations 2.55 and 2.56 are called the *thermodynamic inequalities*. Stability of a system in equilibrium is guaranteed by the validity of such inequalities. States in which such conditions are not satisfied are unstable and cannot exist in nature.

Le Chatelier's principle states that a system in equilibrium, when subjected to variation in one of the factors governing its equilibrium, undergoes a compensating change which tends to moderate the variation. It can also be formulated such that an external influence, disturbing the equilibrium of the subsystem, induces in it processes tending to weaken the effects of this influence. A proof of this principle is presented by Landau and Lifshitz (1959).

Many examples are available which illustrate the validity of this principle. The generalization for stationary states of this principle is evidently important to prescribe the system response behavior to external perturbations.

A moderation theorem given by De Donder is the inequality for a chemical reaction,

$$\mathscr{A}v > 0 \tag{2.57}$$

where $\mathscr{A}$ is the affinity of the chemical reaction and v is the velocity of the reaction. This inequality states that the rate of change of a chemical process and its affinity always have the same sign. For systems comprised of several chemical reactions, it explains the possibility of coupled chemical reactions and at the same time yields a value for the maximum rate of the coupled reaction in terms of the rate of the coupling reaction.

Bak (1963) has shown that stationary states far from equilibrium are stabilized by such a moderation theorem. The fluxes will arrange themselves in such a way as to diminish the perturbation. The concentrations in a system with chemical reactions decay monotonically toward the stationary state. The stationary states far from equilibrium are stabilized in the same manner as near-equilibrium stationary states. The symmetric nature of the matrix of the phenomenological coefficients ensures that the decay toward the stationary state will be monotonic.

In extension of the concept of entropy to nonequilibrium situations, in order to be consistent with the definition of entropy of equilibrium situations, one assumes the limiting conditions of almost vanishing gradients. However, this limits precision of the definition to states of the system very close to equilibrium. The term *entropy, S,* has a statistical or probabilistic meaning, expressed by the famous Boltzmann relation:

$$S = k_B \ln W + \text{a constant} \tag{2.58}$$

W is the probability for the occurrence of a state, k is a constant called the Boltzmann constant, equal to the ratio of the gas constant and Avagadro's number, (R/N). The field of statistical mechanics assigns an exact meaning for the probability of a state and renders a general expression for W in terms of distribution functions or partition functions.

The calculation of entropy for irreversible processes is more complicated. It is evident from the relation

$$dS \geq \{dQ_{rev}/T\} \tag{2.59}$$

where Q_{rev} as the work done in a reversible process cannot be invoked to compute the entropy increase occasioned by an irreversible process. However, it is possible to visualize entropy as a property of the system, capable of flow, from one part of the system to another part. Unlike matter, charge, or energy, *entropy behaves like a fluid, which is not conserved*. One may visualize entropy as a fluid-like property of the system, which may be either created or destroyed within the confines of the system.

The entropy flux density, J_S, is a vector with direction that coincides with the direction of entropy flow and has a magnitude equal to the total entropy crossing per unit area per unit time. The entropy source density, σ_S, and flux density will be functions of both position variable and time.

The entropy balance equation can be written as

$$\{\partial(\rho s)/\partial t\} + \nabla \cdot J_s = \sigma_S \tag{2.60}$$

Equation 2.60 should be compared with the corresponding equation for the conservation of energy,

$$\{\partial(\rho e)/\partial t\} + \nabla \cdot J_e = 0 \tag{2.61}$$

where ρ is mass density and J_e is energy flux density.

When a property under consideration is conserved, the source density of such a property is zero. The conservation of mass is expressed by the equation

$$\{\partial \rho/\partial t\} + \operatorname{div} J_m = 0 \tag{2.62}$$

where J_m is the mass flux density. The mass flux density is definable independently of its balance equation, as the time rate at which matter penetrates per unit area per unit time. The energy flux density, J_e, is defined by its balance equation and can be visualized only as an imaginary fluid.

For a closed system, which has a volume V and entropy S with a surface area a, the volume integration of Equation 2.58 yields

$$(d/dt)\int_V s\, dV + \int_V \operatorname{div} J_s\, dV = \int_V \sigma_s\, dv \tag{2.63}$$

The term in the right-hand side of Equation 2.61 is the entropy produced in volume V per unit time, while the first term in the left-hand side is simply (dS/dt). The second term in the left-hand side can be transformed into a surface integral $\int_a n \cdot J_s\, da$. One has to evaluate this integral using the expression for J_s, on the surface. Since the system is closed, the matter fluxes are zero on the surface, while the energy flux J_e is the pure heat flux density. This results in the statement of the second law for a closed system.

The entropy increase dS of a system can be factored into two terms, namely, the term $d_e S$, which derives from the transfer of heat from external sources across the boundary of the system, and the term $d_i S$, which is due to changes occurring inside the system. According to the second law of thermodynamics, d_iS can never be negative.

$$d_iS \geq 0 \tag{2.64}$$

Equation 2.64 is postulated to be valid for both open and closed systems. If Equation 2.64 is assumed to be valid both for the whole system as well as locally, defining σ_s for each point of the system, one has

$$\sigma_s > 0 \tag{2.65}$$

The assertion that the entropy produced in any process is zero or positive expressed in its global form (Equation 2.64) and in its local form (Equation 2.65) constitutes one of the basic postulates of irreversible thermodynamics.

Since σ is a scalar, various contributions to it should arise from the dot products of two vectors, or the scalar product of two scalars. Thus, if the entropy production is due to a chemical reaction, one has the conclusion that the reaction rate v has the same sign as the affinity ∞. However, this should be understood with caution. The driving force for a chemical reaction is the algebraic difference in chemical potentials of each component of the reacting species and is, therefore, not a vector. The concept of affinity is physically meaningful only when the ratio of this difference in chemical potentials and the product RT, where R is the gas constant and T is temperature in Kelvin scale, is very small compared to unity. This situation is very rare with most chemical reactions.

Within the confines of linear irreversible processes, each flux vector can be deemed to be proportional to its conjugate force, the proportionality constants being measurable and a characteristic property of the substance in question. Thus, one has the empirical laws:

$$J_q = -\text{grad } T, \quad \text{Fourier law}$$

$$J_i = -D_i \text{ grad } \mu_i, \quad \text{diffusion law} \tag{2.66}$$

$$J_E = -\text{grad } \phi, \quad \text{Ohm's law}$$

In the presence of more than one kind of force acting on the system simultaneously, there will be more than one kind of flux present. The magnitude of each kind of flux observed under this circumstance will be different from the value one would observe when the conjugate force of the flux is present alone. For example, when both thermal gradient and concentration gradient are present, the matter flux will arise due to both forces and this may be expressed as

$$J_k = L_{kk} \text{ grad } C_k + L_{kE} \text{ grad } T$$

$$J_E = L_{Ek} \text{ grad } C_k + L_{EE} \text{ grad } T \tag{2.67}$$

Thus, when n processes occur in a system simultaneously, the associated fluxes J_i's are linearly related to the conjugate forces X_i's by the set of relations

$$J_i = \sum_{j=1}^{n} L_{ij} X_j \qquad (2.68a)$$

where the phenomenological coefficients L_{ij} are the partial derivatives

$$L_{ij} = \{\partial J_i / \partial x_j\}|_{X_k; k \neq j} \qquad (2.68b)$$

The reciprocal relations of Onsager states that the matrix of the phenomenological coefficients is symmetric.

$$L_{ij} = L_{ji} \qquad (2.69)$$

The positive definite nature of the rate of entropy production leads to the additional constraint on the phenomenological coefficients. The diagonal elements of the matrix of phenomenological coefficients are positive definite. Although the off-diagonal elements may assume either sign, they must satisfy the inequality that

$$L_{ii} L_{jj} > L_{ij}^2 \qquad (2.70)$$

The symmetry conditions in isotropic systems lead to the validity of Curie's theorem, which forbids the existence of a cross-phenomenological coefficient between forces and fluxes, whose tensorial character differs by an odd number in isotropic systems. This means simply that all coupling terms between vector forces (vectors being a tensor of rank 1) and scalar fluxes (scalar being a tensor of rank zero) are zero in isotropic systems. Thus, there is no phenomenological coupling between diffusion and chemical reactions in solutions.

An elementary proof of this theorem is as follows. The phenomenological coefficient between a scalar flux and a scalar force is evidently a scalar. Similarly, the phenomenological coefficient between a vector flux and a vector flow is also a scalar, purely from dimensional considerations. Therefore, for an arbitrary flux J_1 of tensorial rank n and an arbitrary force X_2 of the same tensorial rank, the phenomenological coefficient will be again a scalar.

Since the phenomenological coefficient L_{12} depends on the properties of the system, namely, temperature, pressure, and composition, it is determined only by the isotropic properties of the system and not by the gradients in the system.

If the force X_2 has a tensorial rank m and the flux J_1 has a tensorial rank of n, $(m - n)$ being an even integer, the tensor product $L_{12} X_2$ is also valid for isotropic systems, since the phenomenological coefficient of even rank is also consistent with the isotropic nature of the fluid system.

However, for the force X_2, differing in tensorial rank n by an odd integer, the corresponding tensorial rank of the phenomenological coefficients will be $(m - n)$ and implies a nonisotropic character for the system. Consequently, such a coefficient must vanish. By definition, an isotropic system cannot have a vector quantity associated with it. Thus, if $(m - n)$ equals unity, the coefficient L_{12} will be a vector. Thus, coupling coefficients between diffusion currents and chemical reactions do not exist in isotropic systems.

The ultimate goal of any physical theory is to provide predictions amenable to experimental verifications. One assumes that in a physical system which is isolated and left for a sufficiently long time, the state will reach ultimately an equilibrium state, where all observable properties remain constant with time. The approach to equilibrium is unidirectional. We learn from statistical mechanics that this state of rest associated with equilibrium is only apparent, and that absolute irreversibility is fiction. With the exception of total mass and total energy, all observable properties in reality are liable to undergo accidental fluctuations, namely, deviations from equilibrium values. Such fluctuations are absent in the total mass and energy for a system which is isolated. Considerations of fluctuations in equilibrium are usually neglected, since one is concerned with measurable and observable properties, and fluctuations decrease very sharply with an increase in the number of particles. Fluctuations are inversely proportional to the square root of the number of particles. Since Avagadro's number is very large, neglect of fluctuations is justified in most macroscopic systems.

Among various possible irreversible processes occurring in a system, one distinguishes what are generally known as stationary states. A system is said to be in a stationary state if its macroscopic parameters, such as temperature, pressure, concentration, and entropy, do not depend on time t, not withstanding the possible occurrence of irreversible processes. The intensive variables defined locally will vary from point to point in the system. Therefore, gradients of scalar quantities will exist. These are the forces which result in fluxes. The forces and fluxes are vectors.

A stationary state represents a stable situation; that is, if a transient interference has caused a small displacement from the stationary state, the system will return of its own accord to the initial stationary-state condition. In the simplest realization of the stationary states, the macroscopic variables have lost their dependence on position also and have become uniform throughout the system. This is the familiar equilibrium situation and constitutes a subclass of the classes of stationary states. Nonequilibrium stationary states can be realized only if the entropy-producing processes are sustained by a continual flow of energy, matter, or both between the system and its surroundings.

Stationary states may be classified as follows. If there are n independent forces and a subset k of these can be kept fixed through the operation of external constraints, the system will sooner or later reach a stage where the remaining $(n - k)$ forces will be constants. Thus, given sufficient time, all forces and their conjugate fluxes become independent of time. Such a stationary state is said to be of the order k, since the additional forces and their conjugate fluxes, namely, $X_{k+1}, X_{k+2}, \ldots, X_n, J_{k+1}, J_{k+2}, \ldots, J_n$, will remain not only as constants but will become identically equal to zero.

Onsager maintained that a deviation from equilibrium which occurs spontaneously in a system as a result of a fluctuation decays on the average in the same manner as a disturbance that has been artificially induced by outside interference. This is known as the *principle of regression of fluctuations*.

NONLINEAR SYSTEM THERMODYNAMICS

The last two decades have seen the development of nonequilibrium, nonlinear thermodynamics in an attempt to bring processes occurring in living systems within the realm of physics and concepts of thermodynamics. The idea of evolution in biology is associated with an increase in organization, giving rise to the creation of more and more complex structures. In thermodynamics, the idea of *evolution implies the most probable state* corresponding to *a state of maximum disorder*. The second law of thermodynamics requires that there exists a state function called entropy, S, which depends on the macroscopic state of the system. For systems which exchange matter and energy with the environment, the variation of the entropy dS during a time t may be factored as

$$dS = d_e S + d_i S \qquad (2.71)$$

where $d_e S$ denotes the *entropy flow* from the surroundings and $d_i S$ represents the *entropy production* within the system, due to the occurrence of irreversible processes. Classical thermodynamics states that isolated systems (that is, systems which cannot exchange matter or energy with their environment) always proceed toward an equilibrium state. During the attainment of equilibrium, the entropy of the system increases monotonically and reaches a maximum value in the equilibrium state. Thus, the content of the second law is

$$d_i S \geq 0 \qquad (2.72a)$$

$$\{d_i S / dt\} \geq 0 \qquad (2.72b)$$

with the equality sign applying only for the equilibrium situations. For an isolated system, by definition, $d_e S$ equals zero. Therefore, one has the familiar form of the second law for an isolated system from Equation 2.71.

The achievements of nonequilibrium thermodynamics that have made it one of the fundamental features of the modern natural sciences are basically due to the heuristic strength of its phenomenological approach. There are a number of important problems which appear to be inaccessible to the phenomenological approach. Irreversibility, which is introduced as a postulate, arises from the reversible character of the laws of motion of individual particles of the system. One should recognize and define the limits of applicability of a thermodynamic description. The assumption that equilibrium exists locally in systems under nonequilibrium conditions, which is essential for the development of nonequilibrium thermodynamics, should be critically examined. If the gradients are large and dispersion is essential, the nonlinear processes never approach equilibrium. In this situation, the local equilibrium hypothesis becomes meaningless.

There also exists an ambiguity in the definition of a nonequilibrium thermodynamic ensemble entropy, and the even greater problem of whether or not the concept of entropy exists for all systems far from equilibrium. For nonlinear situations, the factorization of the entropy production into components from separate identifiable processes becomes questionable. A critical examination of some of these concepts is beyond the scope of this book. Our brief survey of the subject of thermodynamics is concluded at this stage. An account of the recent advances of nonequilibrium nonlinear system thermodynamics is presented in Chapter 8.

Appendix 2A: Proof of reciprocal theorem

If a function $F(x, y, z) = 0$, one has $x = x(y, z)$. Therefore,

$$dx = \{\partial x / \partial y\}_z \, dy + \{\partial x / \partial z\}_y \, dz \tag{2A.1}$$

Similarly, $y = y(x, z)$ and

$$dy = \{\partial y / \partial x\}_z \, dx + \{\partial y / \partial z\}_x \, dz \tag{2A.2}$$

Substitution of 2A.2 in 2A.1 yields

$$
\begin{aligned}
dx &= \{\partial x / \partial y\}_z [\{\partial y / \partial x\}_z \, dx + \{\partial y / \partial z\}_x \, dz] + \{\partial x / \partial y\}_y \, dz \\
&= \{\partial x / \partial y\}_z \{\partial y / \partial x\}_z \, dx \\
&\quad + [\{\partial x / \partial y\}_z \{\partial y / \partial z\}_x + \{\partial x / \partial z\}_y] \, dz
\end{aligned}
\tag{2A.3}
$$

Of the three variables, only two are independent. If x and z are independent, the preceding equations must be true for all sets of dx and dz. If dz equals zero, and dx is not equal to zero,

$$\{\partial x / \partial y\}_z \{\partial y / \partial x\}_z = 1 \tag{2A.4}$$

Thus, one has the reciprocal theorem mentioned in the text. If, $dx = 0$, and dz is not equal to zero, by algebra, one obtains the relation

$$\{\partial x / \partial y\}_z \{\partial y / \partial z\}_x \{\partial z / \partial x\}_y = -1 \tag{2A.5}$$

This concludes the proof of the reciprocal theorem of equilibrium thermodynamics.

Appendix 2B: Maxwell's relations

For a one-component system, the equations of thermodynamics with their corresponding independent variables are

$$dE = T \, dS - p \, dV; \quad S, V \tag{2B.1}$$

$$dH = -S \, dT + V \, dp; \quad S, p \tag{2B.2}$$

$$dF = -S \, dT - p \, dV; \quad T, V \tag{2B.3}$$

$$dG = -S \, dT + V \, dP; \quad T, p \tag{2B.4}$$

Given any exact differential, expressed as $M \, dx + N \, dy$, one can always write

$$\{\partial M / \partial y\}_x = \{\partial N / \partial x\}_y \tag{2B.5}$$

This is the cross-derivative theorem. One has, for example, from Equation 2B.4, the relation

$$(\partial S / \partial p)_T = -(\partial V / \partial T)_p \tag{2B.6}$$

Similarly, from Equations 2B.1, 2B.2, and 2B.3, one obtains

$$\{\partial S / \partial V\}_T = \{\partial p / \partial T\}_V \tag{2B.7}$$

$$\{\partial p / \partial T\}_S = \{\partial S / \partial V\}_p \tag{2B.8}$$

$$\{\partial S / \partial p\}_V = -\{\partial V / \partial T\}_S \tag{2B.9}$$

Equations 2B.5 to 2B.9 are the four well-known Maxwell's equations. These relations are the consequence of the exactness property of the thermodynamic potentials, G, F, H, and E. The left-hand side of all four of Maxwell's equations are derivatives of entropy. Thus, each is proportional to the rate of heat absorption. The right-hand sides are derivatives with respect to temperature. In this manner, Maxwell's equations equate a calorimetric property of a system with a thermometric property of the system.

There are six general relations of Maxwell. The utilization of the Clausius equations 2B.1 to 2B.4 yielded four of these. With the primary set of variables p, V, T, and S, there are two remaining pairs of independent variables, namely T, S. The four relations of Maxwell do not represent four independent statements; they are simply equivalent statements expressed in terms of four different sets of chosen independent variables. No single derivative occurs more than once in these relations. Thus, regarded as algebraic equations in partial derivatives, the four Maxwell's relations are essentially independent. None can be deduced from another by purely algebraic eliminations.

However, using differentiation, any one can be deduced from any of the rest by a change of variables. If for example, one considers Equation 2B.5 and change the independent variable pair from T, p to T, V, one obtains by the reciprocal theorem

$$T, p: \{\partial S / \partial p\}_T = -\{\partial V / \partial T\}_p$$

$$\{\partial S / \partial V\}_T \{\partial V / \partial p\}_T = -\{\partial V / \partial T\}_p$$

$$\{\partial S / \partial V\}_T = -\{\partial V / \partial T\}_p \{\partial p / \partial V\}_T = \{\partial p / \partial T\}_V \quad (2B.10)$$

The last of Equation 2B.10 is just Equation 2B.6. The left-hand side of Equation 2B.5 is the inverse of $\{\partial p / \partial S\}_T$. Using what is known as the composite function theorem, one obtains the relations

$$\{\partial S / \partial p\}_T = -\{\partial V / \partial T\}_p \{\partial T / \partial T\}_p - \{\partial V / \partial S\}_T \{\partial S / \partial T\}_p$$

$$= -\{\partial V / \partial T\}_p - \{\partial V / \partial S\}_T \{\partial S / \partial T\}_p \quad (2B.11)$$

This is the fifth Maxwell's relation, which contains a new relation, involving $\{\partial S / \partial T\}_p$. Similarly, changing to the pair of variables p, V, one obtains the sixth Maxwell's relation,

$$\{\partial S / \partial p\} = -\{\partial V / \partial T\}_p - \{\partial S / \partial V\}_p \{\partial V / \partial p\}_p \quad (2B.12)$$

This sixth Maxwell relation, Equation 2B.12, completes the set, involving the new derivative $\{\partial V / \partial p\}_T$, which is compressibility.

REFERENCES

BAK, T., *Contribution to the Theory of Chemical Kinetics*, New York: W.A. Benjamin Inc., 1963.

FITTS, D.D., *Nonequilibrium Thermodynamics*. New York: McGraw-Hill Book Co., 1962.

KIRKWOOD, J.G., and I. OPPENHEIM, *Chemical Thermodynamics*. New York: McGraw-Hill Book Co., 1961.

LANDAU, L.D., and E.M. LIFSHITZ, *Statistical Mechanics*. Reading, MA: Addison-Wesley Co., 1959.

SYCHEV, V.V., *The Differential Equations of Thermodynamics*. Moscow: Mir Publishers, 1983.

TISZA, L., *Generalized Thermodynamics*. Cambridge, MA: The M.I.T. Press, 1966.

VAIDHYANATHAN, V.S., *Bull. Math. Biology,* vol. 41, 1979, p. 365; *Adv. in Chemistry Series,* #188, "Bioelectrochemistry: ions, surfaces, membranes," ed. M. Blank. Amer. Chem. Society, 1980.

YOURGRAU, W., A. VAN DER MERWE, and G. RAW, *Treatise on Irreversible and Statistical Thermophysics.* New York: The Macmillian Company, 1966.

chapter 3

Features
of Homeostatic
Systems

INTRODUCTORY REMARKS

Homeostasis is the maintenance of a variable that is subject to alterations within a tolerable and usually narrow range of an average value. Regulation may be defined as the sum total of the various processes by which homeostasis is achieved. Body weight, for example, is the typical result of a material balance. Body temperature is the result of a balance between energy input and energy output. The complex structures associated with living systems play a role in the regulation and control of such systems. The high degree of organization in living things results in the modification of one or more of the variable properties of the system which determines the behavior of the system. This in turn initiates a series of events that produces changes in other properties of the system. Many such complex examples are found in the system of chemical reactions catalyzed by enzymes. Living systems contain a large variety of organic molecules, such as proteins, nucleic acids, and carbohydrates in a highly organized manner.

Energy is utilized by living systems to grow, to be mobile, and also to maintain themselves in a stationary state, preventing the system from reaching the equilibrium state. Energy, material, and information pass through the individual animal at various and varying rates during its existence. The ability of animals to maintain constant composition of the extracellular fluid represents one of the most significant advances of evolution, since with this facility, animals became independent of the frequent changes in their environment. A qualitative account of the various information needed to analyze aspects of homeostasis is presented in this chapter.

A remarkable control is exercised over the multitudinous metabolic processes and energy-yielding reactions which proceed in the presence of disturbing conditions of the environment. An organism's ability to maintain a normal constant internal environment, despite numerous metabolic reactions occurring and the changes in the environment, is due to sensitivities of the specific regulatory mechanisms. The rate at which individual reactions should be controlled is in a manner that permits it to proceed at a rate commensurate with the demands of the cell. Biosynthesis of unneeded compounds would result evidently in a waste of energy.

The binding of a ligand, dissimilar to the substrate of an enzyme, at an allosteric site, modulating the catalytic activity of an enzyme, can be pictured on feedback principles. Such ligands induce a conformational change of the enzyme structure which results in either an increase or decrease of the enzyme activity. In this manner, the intracellular concentrations of a particular chemical can regulate the rate of a metabolic pathway in accordance with the needs of the organisms. The final product of the pathway modifies the activity of the enzyme catalyzing the first committed step of the pathway. In many examples, the product of one metabolic sequence modulates the velocity through a separate but related sequence. Such metabolic regulation may be termed *feedback inhibition*. It is an essential mechanism for the maintenance of stationary-state concentrations, since it prevents the unnecessary accumulation of a given metabolite.

Similarly, feedforward activation, namely, the stimulation of enzyme activity by a metabolite, may result in a propagation phenomenon. While negative feedback may correct for an overshoot in response of a pathway to perturbation, positive feedback may result in a regulatory pattern. This insures constancy of the biochemical environment through a more complicated mechanism, which involves initial destabilization.

EFFECT OF MEMBRANES

The membranes of cells of living systems perform dual functions. (1) They maintain the identity of the internal medium of solutions of electrolytes from the external solutions of different compositions. (2) The membranes allow passage of nutrients selectively across the barrier. The permeation mechanisms of various nutrients across the cell membranes play a significant role in homeostatic mechanisms.

Many important aspects of the regulatory phenomena are centered around the plasma membranes. Presumably, the membrane components can undergo configuration shifts more easily than protein molecules. The membrane also serves as a surface for anchoring. An osmotic pressure can arise when two solutions of different concentrations of solutes are separated by a membrane, permeable to some of the species, and impermeable to other species of solutions in contact with the membrane. Osmotic flow continues until the chemical potential of the diffusing components is the same on both sides of the barrier. If the flow can occur into a closed volume, the pressure inside necessarily increases. Osmotic pressure is essentially a manifestation of the thermodynamic requirement of equality of the chemical potential of the solvent, water, on both sides of the membrane.

The quantitative justification of the preceding statements is as follows. If an aqueous solution of r components, in side 1, is separated from the pure liquid, 1, present on side 2, by a semipermeable membrane, under equilibrium conditions, the chemical potential of the solvent and the temperatures should be the same on both sides.

$$T^1 = T^2$$

$$\mu_1(T, p, x_1, \ldots, x_{r-1}) = \mu_1(T, p_0) \tag{3.1}$$

μ_1 is the chemical potential of species of kind 1. p is the pressure in side 1, required to achieve equilibrium between solution of side 1 and the solvent in side 2, and p_o is the pressure in side 2. Using the relation

$$\{\partial \mu_1 / \partial p)_{T,x} = v_1 \tag{3.2}$$

where v_1 is the partial molar volume of solvent. One obtains for the difference in the chemical potential of the solvent on both sides of the membrane as

$$\mu_1^{(2)}(T, p) - \mu_1^{(2)}(T, p_o) = v_1^0 \tag{3.3}$$

For an ideal solution, one obtains the expression for the osmotic pressure, π,

$$\pi = c_2 RT \tag{3.4}$$

where c_2 is the concentration of solute in the solution. Equation 3.4 is called the van't Hoff law.

Similarly, if a membrane (rigid, heat conducting), permeable to some of the components but not to others, separates two electrolyte solutions, for equilibrium, the temperature and the chemical potential of the diffusing species should be equal on both sides. Thus, one has from the expression of the chemical potential of ionic species of kind i, namely,

$$\mu_i = \mu_i^\star + Z_i F^\star \phi$$

where $\mu_i^\star$ is a function of temperature, pressure, and composition alone,

$$\mu_i^{\star(2)} - \mu_i^{\star(1)} = -Z_i F^\star \{\phi^{(2)} - \phi^{(1)}\}$$
$$= -Z_i F^\star \Delta\phi \tag{3.5}$$

where $F^\star$ is the Faraday constant, and $\Delta\phi$ is the difference in electric potential between the two sides of the membrane.

Therefore, if there exists a difference in the electric potential across a membrane which is nonzero, this implies that the nonelectrical part of the chemical potential of the permeable ionic species, i, be unequal on both sides of the membrane. Thus, activities or concentrations of the specified ionic species should be different on both sides. The Nernst equilibrium expression Equation 3.6 for the electric potential difference will be valid if and only if the expression in Equation 2.34 is valid. The validity of Equation 2.34 is restricted to extremely dilute, ideal solutions.

The Donnan membrane phenomena, arising from the existence of impermeable polyvalent ions on one side of a membrane with the existence of other permeable small ions on either sides of the membrane, at equilibrium, is essentially the result of the two opposing tendencies of the electrolyte system to maintain equality of chemical potentials on both sides of the membrane and the need to maintain electroneutrality.

$$\ln\{c_i^{(2)}/c_i^{(1)}\} = (Z_i e/k_B T)[\phi^{(1)} - \phi^{(2)}] \tag{3.6}$$

In Equation 3.6, e denotes the charge of a proton and k_B is the Boltzmann constant. $(e/k_B T)$ equals $(F^\star/RT)$, where R is the gas constant.

These considerations apply strictly only to equilibrium situations. Equilibrium is rarely achieved in living systems. Thus, contributions from velocities of moving ionic species for the electric potential should be included in the computation of stationary-state electric potential differences across real biological membrane systems.

Although regulating systems and controls employed in engineering devices and in biology operate mainly by negative feedback, there are significant differences in the organization of the two categories. The biological homeostatic system is a functioning system that has acquired its precision of operation and complexity over a long period of evolution. Our understanding of how the various components contribute to total system performance is very important.

Molecular interactions constitute the central subject of chemistry. Covalent bonds represent the strongest molecular interactions that sustain components of living organisms. Once the structure of living systems is stabilized by covalent forces, it is the weaker forces that provide for the flexibility of living organisms to respond to environmental changes in conditions. This reliance on noncovalent forces results in a lack of rigidity (that is, flexibility or instability) of the protein structures consequently in protein function.

The intrinsic structures of noncovalent bonds are much more difficult to characterize than covalent structures. Due to their exceptional sensitivity to environmental conditions, they provide the plasticity of response of living systems. Such noncovalent forces are due to electrostatic interactions, van der Waals' forces, and hydrogen bonds.

Living organisms do not usually influence the rate of metabolic reactions by invoking changes in temperature, nor can most survive at high temperatures. Therefore, a catalyzed reaction is necessary to make reactions proceed rapidly. In addition, if biological reactions proceed without catalysis, no control can be exercised over their rates. If the organism loses control of the rates of important vital reactions, the maintenance of normal structure and functions becomes impossible.

An overall biochemical view of regulation may be stated as follows. The biochemical implications of evolution and the functional specialization of molecules like DNA, RNA, and various proteins should be recognized. The modulation and modification of enzyme-influenced chemical reactions and the regulation of the synthesis of various required molecules should be appreciated. The influence of ligand molecules and hormones and their role in DNA replication and the cell cycle as well as the role of membrane-bound enzymes form the focus of biochemical research. The physicochemical aspects of structure and the transport of molecules as well as the role of compartmentalization play the complementary aspects of the description of regulation. The servomechanisms and oscillatory phenomena occurring in biology could be analytically studied by the use of mathematics of differential equations. The concepts of hypercycles and dissipative structures as well as extension of thermodynamics to nonlinear nonequilibrium systems provide an umbrella for overall consistency.

Proteins have the ability to undergo characteristic structural changes from one shape to another. The regulation function of proteins arises from such conformational changes. Proteins with very pronounced regulation based on allosteric transition are usually made up of more than one kind of subunit. The mechanism for allosteric transition can be stated as follows: The addition of the regulatory molecule, as well as of the substrate, involves weak bonds so that there is an equilibrium between attached and free molecules. At a certain concentration of these molecules, the protein is attached with them for a certain period of time. When such times are sufficiently long, a change in the structure of the enzyme is likely to occur. A regulatory subunit connected with a regulatory molecule changes the conformation to one more energetically favorable. This results in a change in structure of the whole protein molecule, which affects the rate of interaction of the enzyme with the substrate. A common result of these changes is that the regulatory molecule causes a certain inhibition of the enzyme reaction at low concentrations of the substrate. This cooperative-type kinetics involves participation of more than one subunit. Regulation by a repressor is negative by preventing a specific synthesis. Our knowledge of these regulatory molecules and their mechanisms is still rudimentary.

ROLE OF CHEMICAL REACTIONS IN REGULATION _____

The synthetic sequences which lead to the formation of the required end products, such as amino acids, purine and pyrimidine nucleosides, and steroids, begin with the simpler cell materials that arise from the metabolism of the carbohydrates or fatty acids. In each metabolic pathway, there occurs a *committed step*, the reaction which produces the first metabolite which has no role in the metabolism other than to serve as an intermediate in the biosynthesis of the end product of the sequence of reactions. In most instances, the committed step proceeds with a loss in free energy so that the reaction is almost irreversible. It is evident that a metabolic control intended to restrict the formation of an end product would function most satisfactorily.

The control of the committed step in the metabolic reaction sequences turns out to be an almost invariant attribute of metabolism; control of intermediate steps is exercised less frequently. Five kinds of regulation of chemical reactions are observed in mammalian systems. These are:

1. Regulation of the rate of entry of nutrients into cells.
2. Repression of enzyme synthesis by an end product of a metabolic sequence.
3. Induction of the formation of an enzyme by a nutrient.
4. Induction of an enzyme reaction.
5. Stimulation by a metabolite of enzymes which function in the metabolic pathways.

The rate of a chemical reaction is dependent on the product of the concentrations of reacting substances with a rate constant characteristic of the reaction. This result is due to the fact that in order for two molecules to combine for the formation of a new product, they must collide. The frequency of such collisions is evidently proportional to the number of each kind of molecules, hence proportional to their respective concentrations. This is the basis of the *law of mass action*. Thus, if one molecule of A combines with one molecule of B, the rate of production of the product P per unit volume is given by

$$\{d[P]/dt\} = K[A][B] \tag{3.7}$$

where the square brackets, [ ], are utilized to denote concentrations. t is a time variable.

If two molecules of A combine with one molecule of B, the rate is now given by

$$\{d[P]/dt\} = K[A]^2[B] \tag{3.8}$$

Similarly, for the case of unimolecular reactions where a single molecule of A decomposes into two or more products, the rate of formation of each product is

$$\{d[P]/dt\} = K[A] \tag{3.9}$$

When the reaction involves an enzyme, the reaction rate is governed by the Michaelis-Menten scheme. Enzyme-catalyzed reactions are believed to involve two steps: (1) the substrate combines with the enzyme to form an intermediate complex, which is usually a fast reaction; and (2) following at a relatively slower rate, the intermediate decomposes into the product and the enzyme. One also assumes that the second step is almost irreversible. The fact that some enzyme is tied up as an enzyme–substrate complex causes the reaction to be not proportional to the total concentration of the enzyme molecules, since the first step is proportional to only the free unbound number of enzyme molecules.

If E^T denotes the total amount of enzyme molecules in terms of concentrations, and E^F is the concentration of free enzyme, then one has

$$E^F = E^T - [ES] \tag{3.10}$$

where $[ES]$ denotes the concentration of the enzyme–substrate complex. Thus, the reactions catalyzed by the enzyme are usually denoted as

$$E^F + [S] \underset{k_2}{\overset{k_1}{\rightleftharpoons}} [ES] \overset{k_3}{\longrightarrow} [P] + E^F \tag{3.11}$$

The rate of production of the enzyme–substrate complex is now given by the equation

$$\{d[ES]/dt\} = k_1[S]E^F - (k_2 + k_3)[ES]$$

$$\{d[P]/dt\} = k_3[ES] \tag{3.12}$$

Substituting Equation 3.10 in Equation 3.12 and asserting that under stationary-state conditions $\{d[ES]/dt\} = 0$, one has

$$[ES] = \{[S]E^T/([S] + K_M)\}$$

$$K_M = [\{k_2 + k_3\}/k_1] \tag{3.13}$$

$$\{d[P]/dt\} = k_3\{[S]E^T/([S] + K_M)\}$$

K_M is called the Michaelis constant of the reaction. At low concentrations of the substrate, one evidently has the result that the rate of product formation approximates as

$$\{dP/dt\} = \{k_3[S]E^T/K_M\} \tag{3.14a}$$

At high concentrations of the substrate, the rate is proportional to the total enzyme available:

$$\{d[P]/dt\} = k_3 E^T \tag{3.14b}$$

Equation 3.14a is the case when the product formation rate is substrate limited, while Equation 3.14b denotes the case when such rate is limited by the total amount of enzyme available.

The importance of the preceding enzyme-catalyzed scheme, from the point of view of control systems, can be described as follows. A cell normally produces the product P at a rate of R_P. To accomplish this, it must produce the substrate at the same rate or faster. Now, if for some reason, the cell needs to turn off the production of P, this can be accomplished in two ways. It can produce a competitor substance that reacts with the substrate and removes the substrate from reacting with the enzyme. To do this, it should produce the competitor molecule at a rate comparable to R_P.

A more economical way of accomplishing this is to produce a competitor that reacts with the enzyme molecule, thereby reducing the available enzyme for the reaction. Since the number of enzyme molecules available is much less, this is more economical. Thus, a very small amount of competitor that reacts with the enzyme can control the rate of production of the product. Often, the competitor is a metabolic product, and the mechanism thus described is called *product inhibition*. Such control is very effective. If a cell is placed in a medium from which a metabolic substance can diffuse into the cell, the cell will often turn off its own production of the substance.

The manner in which product inhibition leads to control in a living organism is illustrated in the following scheme. Consider the following sequence of reactions.

$$A \underset{k_1}{\overset{E_1}{\rightleftharpoons}} B \underset{k_2}{\overset{E_2}{\rightleftharpoons}} C \underset{k_3}{\overset{E_3}{\rightleftharpoons}} D \tag{3.15}$$

where E_i and k_i are, respectively, the enzymes catalyzing the reactions and the appropriate rate constants ($i = 1,2,3$).

Approximately, D is produced at a rate $k_3[C][E_3]$, and C is produced at a rate $k_2[B][E_2]$, assuming that these reactions are substrate limited. If the first step is enzyme limited, B is produced at a rate $k_1[E_1]$.

Therefore, in the stationary state, one has

$$\{d[D]/dt\} = k_3[C][E_3] = k_2[B][E_2] = k_1[E_1] \tag{3.16}$$

If for some reason, the cell needs to turn off the production of D, it can accomplish this by reducing E_3. But the result of this would be to increase the concentration of C, if nothing else changes in the system. The result would, however, be a waste of the cell's energy in increased production of C, which makes it also difficult to reduce production of D.

On the other hand, if C reacts reversibly with E_1, then the amount of E_1 available for the first step is reduced approximately in proportion to the concentration of C. Although this statement is not exact, it is approximately correct for small concentrations of C. Let

$$[E_1] = [E_{10}] - [T_{11}][C]$$

where $[E_{10}]$ is the amount of E_1 available if concentration of C is zero, and T_{11} is a proportionality constant. In this situation, D and C are produced as stated earlier, while B is produced at a rate $k_1\{[E_{10}] - [C]T_{11}\}$.

In the steady state, one has

$$k_1\{[E_{10}] - [C]T_{11}\} = k_3[C][E_{10}]$$

$$[C] = \{k_1[E_{10}]\}/\{k_1 T_{11} + k_3[E_3]\} \qquad (3.17)$$

$$\{d[D]/dt\} = \{k_1[E_{10}]k_3[E_3]/\{k_1 T_{11} + k_3[E_3]\}$$

If $k_1 T_{11}$ is large compared to $k_3[E_3]$, the cell can reduce the production of D by reducing $[E_3]$.

The extremely high catalytic power in conjunction with strict specificity constitutes a most striking feature common to all enzymes. Catalysis requires reduction of activation energies, and the metastable state with its high internal electric field may activate this. Frohlich (1988) has shown that frequency-selective long-range interactions arise from excitation of coherent vibrations, which might then yield a selective attraction for the substrates. The importance of the metastable state for the reduction of activation energies points to the importance of electrostatic interactions within the enzyme substrate complex. Clearly, water and its ions must play an important role in a quantitative analysis of this problem. Living systems are responsive to weak fluctuations arising spontaneously as a consequence of thermal motion.

A prime function of the control mechanisms is to ensure that energy is not wasted and to permit the synthesis of enough metabolite of a specific kind to meet the immediate needs of the cell. These regulatory mechanisms evidently require some kind of communication between the various organs via either the nervous or the circulatory systems. Regulation may also be the simple consequence of the chemical environment of the enzymic systems.

REGULATION OF CELL CYCLE PROCESSES _____

The cell cycle is often considered to be composed of four phases: (1) the gap before DNA replication, G1, (2) the DNA synthesis phase, S, (3) the gap after DNA replication, G2, and (4) the mitotic phase which culminates in the cell division, M. The G1 phase of the cell cycle is a functional period during which cells prepare for the S phase. This is marked by the beginning of DNA, histone, and some enzyme syntheses. G1 was originally defined as a time interval, assigned to the time gap between the readily observed events of mitosis and the DNA synthesis.

During the S phase of the cell cycle, the entire DNA content of the nucleus must be replicated completely and precisely in a period of a few hours. Control of postembryonic cell proliferation occurs before the S phase. Though this formulation is useful, the cell cycle is seen to be more complex upon closer examination. A large number of macromolecular components are assembled, activated, or moved. The existence of dependent relationships is usually not apparent by observing the normal cell cycle. The function of checkpoints in the cell cycle is to ensure the completion of early events before late events begin. Growth and cell proliferation must be regulated to produce an organism having a defined form of coherent structure.

The usual dogma of evolution states that useless excess genes will be eliminated by selection, whereas useful new genes will be maintained. Integration may not have been accomplished by physical linkage of the new gene to its precursor gene. Association of genes into clusters might occur if duplication were followed first by a cistron shift in the genome and then by a crossover between the new and old linkage group in which mutation has occurred. Clusters of genes that govern associated processes will contribute to controlled gene function. Clustering should therefore be preserved on the grounds that control mechanisms for gene function are selected.

The nature of the selection pressure for the appearance and integration of new steps in an enzyme complex may be of many kinds. The value of the product in promoting the synthesis of its own enzyme is an obvious first consideration, since it would provide a positive feedback for the evolution of a complex cellular system. If the enzymatic step is one of a series of energy or molecular precursors that either is stored or used immediately by the cell, the competitive advantage thus secured leads to success. It is possible that enzymatic metabolic pathways could be added on from either end. At the ordered structure (low-entropy) end, the addition of an enzyme might channel in a new source of nutrients, while at the opposite more disordered end, it may provide a more complete breakdown and utilization of a given substrate.

This mechanism of evolution of enzymatic series may apply to the appearance of metabolic cycles. The existence of metabolic cycles which are maintained by condensation of one substrate to another as the first step in the cycle may depend on the retention of similarities of both the substrate and the amino-acid sequence in the enzyme at each step of the pathway.

Organic evolution is always accompanied by an increased complexity in the evolved form. The evolution of each of the many organ systems regulates the cellular environment to a relatively high degree of constancy. All these systems that are associated with water and acid-base balance are control mechanisms that need to operate and should be necessarily of increased complexity.

Wherever a branching pathway occurs, there might be some degree of sequence and structural homology in the enzymes. The mechanism of allosterically controlled metabolic pathways also depends on sequence similarities that may be retained in pathway enzymes. One possibility of the existence of end-product control mechanisms is that an ancestral region that once bound a substrate molecule remains competent to bind an end product derived from the first substrate because of remnants of structural similarities between the two. In the course of evolutionary time, this site that now binds the end product may have shifted in the control enzyme relative to its active site for the first substrate that enters the pathway. The result of this would be the appearance of the capability of the end product interacting with the control-point enzyme in a manner that would alter the effectiveness of the active center to the entering substrate. Activity alteration would be controlled through the end product's capacity to interfere stereochemically or to induce conformational changes in the enzyme. Change of degree of association could affect the activity of the enzyme to the substrate or other controlling molecules.

Genes are duplicated, mutated, transcribed, and translated into enzymes and other proteins and are selected or eliminated over an extended period of time and in the context of other gene-enzyme systems. The apparent capacity of a control mechanism to anticipate is largely a result of integrating the effects of these factors that occur in time into the design of the cell. It can be argued that evolutionary selection pressures operate in favor of systems that procure and channel matter and energy from some source into their duplication and select for systems that regulate the utilization of the matter and energy pool. The two most important control mechanisms are the control of the metabolism and control of gene activity.

Genes can express their activities at all times during the cell cycle in response to specific induction or derepression (response of cells to external stimuli). The synthesis of many enzymes takes place under conditions in which repression is neither maximal nor minimal. The rate at which the cell produces an enzyme is proportional to the rate at which the cell utilizes the end products of the enzyme's action. This regulation is achieved by means of a twin control system of end-product repression of the enzyme synthesis and end-product inhibition of enzyme activity. The degree of repression is in part a reflection of the enzyme's own catalytic activity. Therefore, the regulation of enzyme synthesis under these conditions is to some extent a *self-regulation*. Synthesis under conditions where the enzyme determines its own rate of synthesis in this manner is called *autogenous*. The rate of the production of the enzyme under autogenous conditions is not constant, but periods of rapid synthesis alternate with periods of little net synthesis.

The timing of cell division can be shown to be controlled by the (DNA/mass) ratio in cells growing under special conditions. As in higher organisms, the foundation of cellular evolution is the mutation and selection of genes. These primary unitary changes operate by alteration of a single codon of a cistron, and in turn this change affects the properties of the protein by an alteration in an amino acid. A protein and particularly an enzyme may express the effect of mutation over a wide range of variation in its properties.

Both cyclic and continual classes of biochemical processes occur within cells. The cyclic processes are those that occur at certain definite times of each life cycle, such as the replication of DNA during the S period, while continual processes are those that are more or less persistent, such as ATP production or total protein synthesis. Somatic cells and some embryos possess feedback controls that prevent cells that have not finished DNA replication from entering mitosis. These controls may act on the accumulation or posttranslational activation of cyclin. Presumably, cyclic processes are under close genetic control and are sequentially ordered. Continual processes are needed for viability.

The biochemical machinery that controls the critical events in the life cycle of the cell, namely, cell division, has been identified recently. The genes controlling cell division in simple yeasts have been identified. The proteins regulating the cell division in clam, sea urchin, and frog eggs have also been isolated. The pathway discovered involves two major proteins. One of these is a particular type of enzyme, a kinase, which adds phosphate groups to other proteins. Such phosphate additions and removals are commonly used by cells to modify protein activities. Kinase activation is a signal that sets in motion the cellular changes necessary for division. The concentration of cyclin, the other major protein in the cell cycle, fluctuates increasing just before cells divide and turning on the kinase. Cyclin concentration drops precipitously and the kinase activity subsides allowing the cells to complete the division cycle.

Any regulatory mechanism describing cell growth and division must account for the periodicity of the cyclic processes and relate these to the essentials of noncyclic processes occurring. The continual reactions may be largely regulated by classical mechanisms through the laws of chemical kinetics and equilibria. On the other hand, the temporal control of cyclic processes is very likely affected through sequence of transcriptions and translations, which provide necessary pathways from gene to functional polypeptide. The existence of metabolic cycles which are maintained by the condensation of one substrate to another as a first step in the cycle may depend on the retention of the similarities of both the substrate and the amino-acid sequence in the enzymes at each step of the pathway.

A fundamental general concept that should prove useful is that regulatory constituents are in a dynamic stationary state. The rapid, large changes in internal molecules effected by modest alterations of external factors, as contrasted to the sluggish responses of stable molecules, show that the steady-state mechanism is particularly suitable for control. Production of a regulatory factor ideally should be highly responsive to external stimuli. A fine-tuned response can be obtained if the molecule is dynamically turning over, as are many substances both large and small. Balance of synthesis and degradation then determines a steady-state level that is rapidly changed if either rate is altered. Quantities of RNAs and proteins are determined by their rates of both synthesis and degradation, and amounts of growth-factor receptors are determined by their rates of production and downregulation. The evolutionary invention of allosteric control mechanisms such as end-product inhibitions can be attributed to the basic selection factors, namely, the selective advantage achieved by selection of control mechanisms that assist the cell or organism in stationary-state operations.

There is little doubt that the ability of a cultured mammalian cell to traverse its life cycle is dependent upon a sequence of transcriptions and translations. The degree to which these events are coupled has been investigated with the result that cells were found to be unable to traverse their life cycle when protein synthesis is inhibited. The experimental results indicate that cells cannot cross a transcriptional marker when translation is inhibited. This is consistent with proposed models for control at translation rather than at the transcription level. The sequence should be regarded as a fundamental property and traverse of the life cycle is rigidly impressed in a tightly coupled transcription-translation sequence with continuous translation operating as a major control feature.

Proliferation is defined as the increase in cell number resulting from completion of the cell cycle, as contrasted to growth, which is the increase in cell mass. Cell proliferation is a tightly controlled process in higher organisms. Cells can remain quiescent for long times or can increase rapidly in number. A cell population in vivo proliferates at a rate dependent on the fraction of its cells that are being cycled rather than on the cycling time. Cycling time has a constant duration independent of external conditions. Extracellular factors determine whether a quiescent cell will begin to proliferate and also whether a normal proliferating cell in G1 will continue to cycle or will revert to quiescence.

Cell-cycle events become largely independent of extracellular factors after cells enter into the S phase, where they will go on to divide and produce two daughter cells. These later processes, such as mitosis, depend on intracellularly triggered controls. In cancer, the control of proliferation is deranged. Regulation of cell proliferation is defective in cancer cells, which replicate in vivo at incorrect times and in incorrect locations in the body.

Physiological control of growth initiation is external; it is increased by other cells as required in multicellular organisms. External controls switch the intracellular machinery between quiescent and G1. Extracellular factors affect single-cell organisms in a different way, but control is exerted before DNA synthesis. In higher organisms, the foundation of cellular evolution is the mutation and selection of genes. These primary unitary changes operate by alteration of a single codon of a cistron and in turn this change affects the properties of the protein by alteration in an amino acid. Since the proteins consist of a rather large number of amino acids, the number of single mutations that will alter a protein can be very large. A protein, and in particular an enzyme, may thus express the effects of mutation over a wide range of variations in its properties. These may range from minute changes in its properties to drastic changes expressed in a region crucial to its activity. Thus, variations resulting from a single mutation provide a continuum of properties upon which evolutionary selection pressures may be brought to bear.

The capacity of a species to pace its expansion at such a rate that a low-average rate of energy expenditure can be drawn for a maximum period of time is a characteristic of considerable evolutionary advantage. This characteristic is linked directly to the existence of control mechanisms, and such control mechanisms are genetic characters that provide an organism with a maximum evolutionary advantage. In this respect, they give the appearance of having the capacity to anticipate.

The primary level of regulation of growth and differentiation most likely occurs at the interaction of the genetic and enzymatic components of the cells. A knowledge of the overall composition of a cell and an outline of the regulation of its individual components do not alone provide a solution to the problem of what makes a cell a growing and self-replicating unit. The problem is one of understanding the mutual control and integration of the syntheses of a large number of different components. In order to understand the integration of subcellular processes, it is necessary to have some knowledge of the temporal sequence of events during a single cell cycle. The evolution of enzymes, metabolic pathways, cycles of reactions, and allosteric control mechanisms appear to be a consequence of some thermodynamic constraints. There are several considerations with respect to the enzyme-catalyzed reactions and metabolic pathways.

1. It is the existence of enzymes as such that provides a considerable measure of control. The upper limit of the rate of substrate conversion is controlled by the number of active sites available. This implies that at a fixed level of enzyme concentration, rates are controlled within a broad set point, that is, from zero to the specific turnover number per molecule.

2. The metabolic pathways are also control mechanisms that aid in the achievement of stationary-state behavior, for it is seen readily that the extension of a pathway not only gives rise to a greater energetic yield in a given environment but also adds specificity to the ways in which energy is utilized. Enzymes and metabolic pathways contribute to the controlled use of energies and play a role in the maintenance of a stationary-state behavior. The capacity of a species to pace its expansion at such a rate that a low-average rate of energy expenditure can be drawn for a maximum period of time is a characteristic of considerable evolutionary advantage. This characteristic is related to the existence of control mechanisms, and such control mechanisms are genetic characters that provide an organism with a maximum evolutionary advantage. They give the appearance of having the capacity to anticipate.

3. Metabolic cycles can also be shown to contribute to steady-state operations and would therefore be favored evolutionally. If one inspects such cycles devoid of side reactions, it is evident that they are rate limited by the availability of end products for recycling. Cycles viewed in this manner are analogous to simple mechanical escapements (governors) used to maintain the time constant in clocks.

4. In the presence of side chemical reaction pathways, metabolic cycles are subject to the controlling influence of addition or removal of pathway intermediates. These kinds of controls are significant in regulation.

There are a number of characteristics of a control system that describes how well it functions. If the system is disturbed in some manner, how well is it able to compensate for the disturbance? Over what range is control possible? How long does it take to restore the perturbed variable to its proper and desirable value? Is the control system stable, or can it in some situations overcorrect, thereby producing a worse error than the one it is attempting to correct? Can the system get into an oscillatory condition? The answers to these questions should evidently come from experiments.

SIGNIFICANCE OF WATER IN REGULATION ⎯⎯⎯⎯⎯⎯⎯⎯⎯⎯

Life processes occur in aqueous systems so that water as a unique solvent can function for various metabolites. Condensation reactions often occurring during the formation of biopolymers involve loss of water from the reactants. In terms of regulation, the biological significance of water should be fully appreciated. Although water is the major constituent of most biochemical systems, living cells and tissues are far from being ordinary aqueous solutions. They are extremely inhomogeneous and highly organized systems. At both the microscopic and macroscopic levels, water is uniquely equipped to serve an important biological function. Hydrophobic interactions are associated with the ordering of the structure of water and are often referred to as entropy driven.

Special mechanisms have been developed to regulate the composition of the extracellular fluids in higher animals. Since living cells always consume and produce substances in solutions, there is a continuous traffic inward and outward in response to gradients of concentrations which extend away and beyond the membrane phase on both sides. The amounts of water in the body are increased as a result of chemical reactions in which one of the products is water, and by intake of water, and by water contained in the food materials. There appears to exist a linear function between the stimulus arising from the hypothalamus and the drinking response. Normal drinking occurs several times a day and a drinking center must initiate and control the complex integrated act of drinking. The rate of drinking falls off as loss of water in the body is replaced.

In the regulation of the water content of the body, the kidney plays a vital role. The functional unit of the kidney is the nephron. Each kidney of an adult human has about a million nephrons. Since current physiological knowledge of the kidney and its function in regulation is available in many books of physiology, the details of kidney physiology and its role in regulation are not presented in this book. Special mechanisms have been developed to regulate the composition of extracellular fluids in higher animals. Since living systems continually consume and produce substances in solutions, there is an exchange of nutrients and water across the membranes of the cells. The movement of water between blood plasma and the lymph, which is an important problem in physiology, is assumed to be determined by the combined effects of hydrostatic pressure existing between the plasma and lymph phase and the osmotic pressure difference.

The amount of water in the body is increased as a result of chemical reactions in which one of the products is water, and also by the intake of water. It has been proposed that the release of ordered water molecules during a chemical reaction serves as a potent driving force for successfully completing an otherwise energetically unfavorable chemical reaction. The conformation of biopolymers is maintained by the favorable entropy change occurring during hydrophobic interactions with water molecules around apolar groups, displaced from their ordered patterns to a more probable, less ordered pattern.

Many major disturbances of electrolytes and fluid balance originate in the extracellular fluid. For the functional process of life to occur, water is necessary. The concentrations of water within a cell must be maintained within certain limits for the efficient functioning of the cell's machinery. This is accomplished to a large extent by the development of selective permeability mechanisms of the membranes that surround the cell. The distribution of water between the cell and the internal environment of the body is controlled by the effective osmotic pressure across the cell membrane, which in turn depends mainly on the concentration of solutes in the extracellular fluids. An analysis of the body's water content and water regulation is presented by Reeve and Kulhanek (1967) on the basis of a model and equations derived for the flow of water from knowledge of physiology.

Water has very anomalous properties. It is reasonable to suppose that most of the thermoregulation and control activities of biological systems rest intimately on the properties of water as the major component. The structure of water is also interesting. There are eight outer electrons, but only four are involved in electronc pair formation with the two hydrogen atoms. The shape of water is that of an isosceles triangle. The H—O—H angle is not far from 105 degrees. The strong attraction of the oxygen nucleus tends to draw electrons away from the protons, thus resulting in a net positive charge around the hydrogen atoms. The two pairs of unshared electrons tend to concentrate in the directions pointing away from the O—H bonds. The water molecule has an electrically polar structure.

Theories of water structures have been developing along three main lines: (1) interstitial theories, based on broken down ice structure, with the cavities in the open hexagonal structure partly filled by interstitials (clathrate structures), (2) two-state theories, treating water as a mixture of an open ice-like state and a closer-packed state, and (3) cluster theories, assuming the existence of rather large clusters of four-coordinated water molecules, together with water molecules in less bonded and nonbonded states (Forslind, 1971). All the theories of water are too simple to give a completely correct picture.

In biological systems, the question arises as to what extent the structure of water is altered by the presence of biological macromolecules (proteins) and what consequences such influences may have. Many aspects of control and regulation in biology rely on the influence of macromolecular structures and surfaces on the configuration of the assembly of water molecules. Adjacent to the hydrophilic surfaces and around nonpolar groups, the structural fit of macromolecules to water is important. The role of water structure near nonpolar groups is essential for the stability of protein conformation. Ion-nucleating properties of many steroids and hormone molecules as well as the structural fit to water of a number of biomacromolecules give a strong indication that water structure is involved in many biological functions, which in the end contribute to the regulation activities in biological systems.

The separate solvation of nonpolar groups results in lower entropy than the solvation of a pair or aggregate nonpolar groups tends to bind to each other. This hydrophobic bonding is a more important stabilizing factor for the native protein conformation than intramolecular hydrogen bonds, such as occurs in α-helix and γ-pleated sheet structures. The enthalpy change for the formation of internal hydrogen bonds is less than 1 kcal/mole and does not contribute to stability. When ice melts, the highly coordinated crystalline structure breaks in many places. The heat of fusion of 1.44 kcal/mole indicates that only about 15 percent of such bonds are broken. Denaturing substances such as urea influence the breaking of such hydrophobic bonds. One assumes that the entropy change for cavity formation in water depends on the radius of curvature of the required cavity, being negative for radii less than 3.5 angstroms, and positive for flat surfaces. Near a flat hydrophobic surface the enthalpy may be lower, owing to van der Waals' attraction, but the surface entropy is expected to remain positive.

A very different type of structure may be stabilized by surfaces that can form or accept hydrogen bonds to water. The importance of this aspect is that if macromolecular surfaces for arrays of hydrogen bonding sites fit to a regular water lattice, as in ice, such structures may become stabilized in the liquid. Biological molecules may fit to water structures. The appropriate values of distances and arrangements on the surface have a clear role in the molecular aspects of regulation.

The specific behavior of sodium and potassium ions observed in biology may also be due to the properties of ordered water. The hydrated ion radii of various alkali metal ions are different from the sizes of naked ions. In transport properties, combinations of self-diffusion coefficients occur and this may also be an origin of specificity (Vaidhyanathan, 1966). With ordered water structures in interfacial regions, the sodium and potassium ions may have very altered mobilities. Potassium ions with their 1.3-angstrom radius could fit much better in a water lattice than sodium ions can.

An interesting proposal of possible polypeptide and protein structures fitting to a water lattice has been presented by Warner (1961). Making use of the cis-type peptide linkage, it appears to be possible to fold a polypeptide chain in a planar arrangement with all the carbonyl oxygen atoms on one side in a hexagonal pattern and all side chains turned on the other side. The distance between carbonyl oxygens in the hexagonal pattern is 4.8 angstroms, fitting the expanded ice lattice. Berendsen (1967) has discussed the importance of Warner's models. Warner's model may or may not represent adequately aspects of protein conformation in solution. But the proposed models are interesting with respect to biological activity. The interesting feature of Warner's models is the fit to the water structure in a hexagonal array. Such a surface influences the water structure in the sense that ice-like structures will be stabilized near the surface, and long-range ordering parallel to the surface will occur in a thin layer. Because of the 4.8-angstrom distance occurring, the structure is likely to be stabilized by the protein-hydrophilic surface. This implies that entropy and enthalpy values, although lower than in free water, remain higher than with ice structures, since the configurational entropy due to the distribution of interstitials themselves do not vanish.

In a group of water molecules clustered together, a positively charged region in one of the molecules tends to orient toward a negatively charged region in one of its neighbors. From the nature of the electrical attractions involved, these hydrogen bonds are collinear, with O—H $\cdots$ O distance of about 2.76 angstroms. The distance from a proton to the nearest oxygen of a neighbor molecule is about 1.77 angstroms. Hydrogen bonds are weak compared to true covalent bonds, such as O—H linkages in a single water molecule. The average energy required to break such covalent bonds is of the order of 110 kcal/mole of such bonds, while it requires only about 12 kcal/mole or less (the lower bond may be 4 kcal/mole) to cleave hydrogen bonds.

The two structural features which cause liquid water to be a highly coordinated type of system with strong interactions are:

1. The water molecule is electrically a highly polar structure, and this polarity is manifested through the formation of hydrogen bonds.

2. The number of protons which form positive ends of the hydrogen bonds around any given oxygen atom is equal to the number of unshared electron pairs which form the negative ends. Thus, the biological significance of the properties of water is one of the most important aspects that should be taken into account in any consideration of regulation and control in biological systems.

The high heat capacity of water is of primary importance in thermal regulation. The specific heat of water, namely, the heat capacity per gram is by definition unity, which is larger than the heat capacities of other liquids such as alcohol or hydrocarbons. Only for liquid ammonia, the heat capacity is about 1.23. The larger the specific heat, the smaller is the temperature rise produced by a given amount of heat energy. Thus, the problem of temperature regulation is far simpler in systems operated chiefly in water than in any other liquids.

The heat of vaporization of water is also very high compared to values of other substances. It is about 595.9 cal/gram for water, while it is of the order of about 100 cal/gram for most other organic liquids. This property of water stabilizes not only the living organism but also its environment. The vast energy absorbed from the sun by the oceans and other bodies of water is converted into the latent heat of evaporation in hot climates, thereby reducing the rise in temperature to a minimum. In colder regions, this latent heat is released back as condensation.

The latent heat of the fusion of water is not of much interest in biology, since the freezing of water inside a living organism is a much rarer phenomena. However, the expansion of water in freezing is of significant biological interest. The density of water is at a maximum at 4 degrees Centigrade, and ice is less dense than liquid water. If ice were heavier than water, it would sink to the bottom of lakes and oceans on freezing. Thus, sunlight in this case would not be able to melt the ice. The less dense property of ice is, therefore, an essential requirement for the regulation of earth surface temperature as well as the survival of many fish and other living creatures of the sea.

Water also has the highest surface tension of any known liquid, except for certain metals or fused salts, which are not of any significance with reference to biological problems. When the surface of a liquid is increased, molecules which were formerly in the interior must be brought to the surface. Work must be done against the attractive forces which operate between the molecules and their neighbors. These forces are the same as those against which work must be performed in vaporizing the liquid. Thus, it is not surprising that substances with high heat capacity have in general a high surface energy. This high value of surface tension is responsible for the rise of water in capillary spaces of soil and plants, although it is inadequate to explain the rise of water in tall trees where osmotic and other forces play a significant role.

The dielectric constant of water far exceeds that of any other pure liquid, except hydrogen cyanide or farmamide. The electrostatic interactions between two charged species in solution are therefore reduced to a much smaller value, when the dielectric coefficient of solvent is much greater than unity. The implication of the high value of the dielectric coefficient of water, approximating 80, is very important in biological systems, since water is capable of dissolving a greater variety of salts, acids, and bases than any other medium.

The electric field around an ion is very large and, therefore, the orienting force is very great. At a distance of 10 angstroms from an univalent ion, the field can be calculated to be of the order of 0.1 million volts/cm, when the dielectric coefficient of intervening space is of the order of 80. Clusters of water molecules congregate and are oriented around an ion in solution. The intense electric field causes the ions of opposite charges to pack very closely. The oriented shells of water molecules around ions produce electric fields of their own, in a direction opposite to that of the field caused by central ions, thus reducing the value of interionic forces. Ions are present in all biological fluid systems and exert a profound effect on many biological functions. Therefore, the importance of the high value of the dielectric coefficient of water becomes evident in regulation and control.

Ion-solvent interactions are analyzed theoretically by the concept of free energy of solvation. *Free energy of solvation* is defined as the change in energy undergone by an ion in passing from a point in a vacuum infinitely far from the solvent into the bulk of the solvent. The real free energy of ionic solvation is defined as the total transfer energy, which may be split into two parts: an electrostatic contribution owing to the passage of an ion through the surface potential at the vacuum-solvent interface and the free energy of the ion-solvent interaction or chemical solvation.

The attempts to calculate the free energy of solvation in a given solvent have been only partially successful and were mostly directed toward the calculation of ionic contributions. In the Born model, the ion is represented by a rigid conducting sphere of radius, r, with the specific charge $Z_\sigma e$, (Z_σ being signed valence charge number of ions of kind σ), in a medium of dielectric coefficient ε.

The change in free energy is experienced when the sphere is transferred from a vacuum into the solvent. The free energy of solvation ΔG of ion is expressed as

$$\Delta G = -\{(Z_\sigma e)^2/(2r\varepsilon)\}[\varepsilon - 1] \tag{3.18}$$

This expression assumes that the ion has identical radii in the gas and solution phases. The neglect of dielectric saturation caused in the medium by the intense field of the ions does not improve the correlation between experimental and calculated values.

A more rigorous approach is to utilize the fundamental equation for the free energy of a dielectric continuum in an electrostatic field:

$$G = (1/4\pi) \int E \cdot dD \, dv \tag{3.19}$$

where E and D are the field strength and electric displacement vectors, in a volume element dv. If the dielectric coefficient is assumed to be independent of the field strength, which is rarely the situation, Equation 3.19 becomes equivalent to Equation 3.18.

GENERAL CONSIDERATION OF PERMEABILITY

In order to describe flow across the membranes in a phenomenological manner, the flows J_i and forces X_i should be defined. If the flows occur along a direction x, perpendicular to the xy plane of the membrane, the driving force per mole of the substance is the gradient of its chemical potential:

$$X_i = -\{d\mu_i/dx\} \tag{3.20}$$

This causes a flux,

$$J_i = C_i v_i \tag{3.21}$$

where v_i, the velocity of species along the x-direction, and C_i is the local concentration of the species (moles/unit volume).

In general, the chemical potential, concentrations, and velocities of species i in the membrane phase varies with the position variable x. From irreversible thermodynamics, one has the linear phenomenological relation

$$J_i = \sum_j L_{ij} \{ -d\mu_j(x)/dx \} \tag{3.22}$$

Since local values of the chemical potential or its gradient cannot be measured, Katchalsky and Curran (1967) suggested integration of the gradient of chemical potential over the dimension of the membrane and wrote

$$\Delta \mu_i = \sum_j J_j R_{ij}$$
$$\tag{3.23}$$
$$J_i = \sum_j L_{ij} \Delta \mu_j$$

R_{ij} are again phenomenological coefficients, relating fluxes with forces. The matrix of coefficients R_{ij} are the inverse of the nonsingular matrix with coefficients L_{ij}. In Equation 3.23 the difference in chemical potential of a species i on either side of the membrane is identified as the thermodynamic force responsible for the fluxes. This difference is a scalar. (The dimension of chemical potential is energy per mole. The dimensions of matter flux, concentrations, and velocity are respectively, moles/unit area/unit time, moles/unit volume, and length/unit time, which are all consistent with dimensional considerations of Equation 3.21. The dimension of the dissipation function, which should have the dimension of the product of matter flux and gradient of chemical potential, is thus energy per unit time per unit volume. Equation 3.23 implies that the dimension of matter flux is moles/unit volume/unit time, which is an inconsistency to be recognized.)

It is convenient to convert the expression for the chemical potential gradient in terms of other suitable measurable quantities. These are the differences in mechanical pressure on both sides of the membrane, $\Delta p = p^I - p^{II}$, and the difference in osmolar concentrations of ith species, $\Delta C_i = C_i^I - C_i^{II}$, and in terms of the difference in electric potential between the two sides of the membrane $\Delta \phi$.

Since the chemical potential difference can be written as

$$\Delta \mu_i = V_i \, \Delta p + RT \, \Delta \ln C_i + Z_i e \, \Delta \phi \tag{3.24}$$

where V_i is the partial molar volume of species i, for nonelectrolytes, when $\Delta\phi$ term is absent, one may write for a binary system consisting of a solute s and solvent w,

$$\begin{aligned}
\Delta\mu_w &= V_w\,\Delta p + RT\,\Delta\ln C_w \\
&= V_w\,\Delta p - RT\{\Delta C_s/C_w\} \\
&= V_w\,\Delta p - \{\Delta\Pi/C_w\}
\end{aligned} \tag{3.25}$$

$$\begin{aligned}
\Delta\mu_s &= V_s\,\Delta p + RT\,\Delta\ln C_s \\
&= V_s\,\Delta p + RT\{\Delta C_s/C_s\} \\
&= V_s\,\Delta p + \{\Delta\Pi/C_s\}
\end{aligned} \tag{3.26}$$

It should be recognized clearly that Equation 3.24 is only an approximation and is valid only for an infinitely dilute solution. This approximation is adopted again in the derivations of Equations 3.25 and 3.26 in the identification of the osmotic pressure difference. Denoting the conjugate flux of the pressure difference as J_V and the conjugate flux due to osmotic pressure difference as the diffusion flow J_D, the dissipation function per unit area of the membrane is written as

$$\sigma = T\{d_i S/dt\} = J_V\,\Delta p + J_D\,\Delta\Pi \tag{3.27}$$

The phenomenological equations for the volume and diffusion flows can be written as

$$\begin{aligned}
J_V &= L_p\,\Delta p + L_{pD}\,\Delta\Pi \\
J_D &= L_{Dp}\,\Delta p + L_D\,\Delta\Pi
\end{aligned} \tag{3.28}$$

L_p represents the volume flow J_V per unit pressure difference Δp when concentrations of solute are equal on both sides of the membrane. This is therefore identified as the mechanical coefficient of filtration across the membrane. The coefficient L_D defines the exchange flow J_D, when pressures on both sides are equal, and there is a concentration difference. The cross–phenomenological coefficients L_{pD} and L_{Dp} obey the reciprocal relations and are hence equal to each other. The ratio of $-L_{pD}$ to L_p is called the reflection coefficient, denoted by $r^\star$.

$$0 < r^\star = -\{L_{pD}/L_p\} < 1 \tag{3.29}$$

From Equations 3.28 and 3.29, one obtains

$$J_V = L_p\{\Delta p - r^\star\,\Delta\Pi\} \tag{3.30}$$

Katchalsky pointed out that Equation 3.30 with $r^\star$ equal to unity has been utilized by physiologists for a long time. Since Equations 3.29 and 3.30

involve easily measurable quantities, such as hydrolytic pressure differences and the osmotic pressure difference computed using values of measurable concentrations of solutes, this phenomenological approach has attained great popularity among experimentalists.

Del Castillo and Mason (1980)† have commented that a partially sieving membrane cannot be made to behave like a semipermeable membrane by manipulation of the coefficients in transport equations, unless the viscous flow terms are discarded. In the derivation of the phenomenological coefficients for J_V and J_D, viscous flow has been omitted from the very beginning.

The variables most easily experimentally studied are the chemical concentrations of the constituents in the blood. The constituents could be the respiratory, ionic, and hormonal constituents. Hemoglobin is a buffer system which accepts protons and mitigates any pH changes, as carbon dioxide diffuses into blood from the tissues. This simple physicochemical buffering is further supplemented by the so-called Bohr effect, by which oxyhemoglobin is a stronger acid than reduced hemoglobin, and oxygen is driven off from hemoglobin by an increase of carbon dioxide or hydrogen-ion concentration.

In maintenance of normal health, the regulation mechanisms of electrolyte balance and acid-base balance are important. Many major disturbances of electrolyte and fluid balance originate in the extracellular fluid. Under usual environmental conditions, there is a loss of approximately 1,500 ml of water by normal human adults. Of this about 600 ml is lost through the skin as insensible perspiration. Any excess intake of water over this obligatory total volume appears as an increased urine volume. To the extent to which the intake is less than 1,500 ml of water, the difference must be at the expense of total body water. The electrolyte composition of the intracellular fluid differs from the electrolyte composition of the surrounding interstitial (extracellular) fluids markedly. The marked disparity in composition of the fluids on either side of the cell membranes cannot be attributed solely to the permeability characteristics of the membrane. Very often, solute transport is accomplished by drag caused by water flow.

---

† Del Castillo and Mason (1980) show that the transport equations for leaky and open membranes have exactly the same mathematical form, differing only in the magnitudes of some of the transport coefficients. Open membranes allow all components of a mixture to pass through, whereas semipermeable membranes strictly prevent the passage of some components. It is impossible to obtain statistical mechanical transport equations for a leaky membrane to pass smoothly to the corresponding equations for a perfectly semipermeable membrane, in contrast to the manner that some phenomenological coefficients behave by manipulation of the coefficient $r^\star$. The equations are not the same for the two cases, since the equilibrium states for the leaky membrane and a semipermeable membrane are fundamentally different.

ACID-BASE REGULATION

Mammalian organisms appear to regulate the pH (related to hydrogen–ion concentration) of their extracellular fluids with precision. The mean normal blood of a healthy individual has a pH of about 7.4 ± 0.05. In disease, the pH of the blood varies more widely than this value. The extreme values compatible with life are about 7.00 and 7.80. Buffers provide the first line of defense against acid–base disturbances. Basically, two types of mechanisms are regarded as controlling blood pH. The first one is the physicochemical based on buffer action of the blood proteins and chemicals in the body fluids. The second set of mechanisms is referred to as physiological mechanisms concerned primarily with the functions of the roles played by the lungs and kidneys and their effects on the buffers of the blood.

A buffer is a mixture of a weak acid and its alkali metal salt, or a weak base together with its acid salt. A mixture of this kind resists changes in hydrogen–ion concentration of the solution upon addition of either a strong acid or a strong base. The chemical mechanism by which buffers function in resisting a pH change is easily understood on the basis of the law of mass action. The pH of a buffer solution composed of a weak acid and its alkali salt can be calculated with the use of the Henderson-Hasselbalch equation:

$$pH = pK' + \log\{[salt]/[acid]\} \tag{3.31}$$

where pK' is a constant for the buffer at a given concentration and temperature, and equals the pH of the solution when molecular concentrations of salt and acid are equal.

The bicarbonate buffer systems of both plasma and erythrocytes play a special role in the regulation of blood pH. The bicarbonate-carbon dioxide system is a very good buffer in vivo, due to the fact that the weak acid member of the pair, H_2CO_3, is volatile as well as it is available in large amounts as a product of oxidative metabolism.

THE CONCEPT OF ACID-BASE BALANCE

The concept according to Bronsted-Lowry is that acids can be defined as substances that donate protons and bases are substances that accept protons. The ionization constant for the first dissociation of carbonic acid can be written as

$$K_a = \{[H^+][HCO_3^-]/[H_2CO_3]\} \tag{3.32}$$

Therefore, acid–base balance refers to the hydrogen-ion concentration regulation as determined by the chemical balance of acids and bases.

In body fluids, the bicarbonate-carbonic acid system represents a principal buffer system. The pK' of this system has a value in blood plasma of 6.10 at 38°C. This buffer system is maintained in the presence of a large and continuous metabolic production, transport, and respiratory excretion of carbon dioxide. The ratio of carbonic acid and bicarbonate concentrations at pH 7.4 is about 20:1. This ratio is determined by renal capacity to adjust normal concentrations of the anions and cations (which are mainly chloride and sodium ions) in the extracellular fluids and thus maintain proper concentration of the carbonic acid, and by pulmonary excretion of carbon dioxide to maintain proper bicarbonate concentrations.

The two factors subject to the action of regulatory and metabolic activities of the body are the whole blood buffer anion concentration and the arterial or alveolar carbon dioxide pressure. The primary changes in the hydrogen-ion concentration homeostasis as reflected in the blood are acid-base balance, and are related to two major pathways of disturbances (Muntwyler, 1968). Kidneys are programmed to excrete excess hydrogen ions in defense of constant hydrogen-ion concentrations of the body fluids. The removal of H^+ ions from the blood can occur by the ingestion of buffer salts, or by the actual loss of HCl through vomiting.

In some respects, the removal of excess water and electrolytes offer greater problems than replacement of deficits. Frequently, patients with edema or heart failure exhibit lowered plasma sodium concentrations. This appears as a derangement of the volume regulation and osmoreceptors controlling ADH and aldosterone elaboration. If one attempts, in these cases, to raise the plasma sodium concentration to normal by restricting body fluid intake, or by administering salt, marked thirst results, and patients exhibit discomfort.

In general, a negative salt balance can be achieved by restricting salt intake, increasing the renal or extra-renal losses of salt. Large amounts of water move through the gastrointestinal tract each day, and almost all of this water is reabsorbed. The loss of water from the body through the kidneys, again responding to osmotic forces, is discussed thoroughly in other books on this subject. Control of the excretion of water by the kidney is influenced by feedback mechanisms. Water intake is controlled by the central nervous system and is usually activated by a deficit in the body water content. The receptors that provide this information for the control system include those sensitive to changes in effective osmotic pressure and probably those sensitive to temperature.

The cell has two basic control systems. One is the control system for growth and multiplication; the other is for function. The two regulation systems are interconnected. A cell operates in the following manner. There is a repository for information, the DNA molecules, which are able to replicate to make more DNA. DNA can transfer its information to molecules of a different kind, RNA, which are smaller molecules. The RNA will transfer its information to proteins, which are the basic functional constituents of the cell.

Regulation occurs through interactions between various cell constituents. The interaction of proteins with nucleic acids can be specific. Transfer of information between the nucleic acids and proteins involves a small number of elements. In contrast, the transfer of information between nucleic acids and proteins requires a much larger surface of interaction. Among the protein molecules that interact with the DNA, there is an important class, called the repressor molecules, whose function is regulation. A repressor recognizes certain DNA sequences located at the beginning of the gene. There is regulation of the synthesis of DNA on RNA, and of RNA on DNA; it is also likely that there are mechanisms which regulate the synthesis of protein on RNA. The effect of regulation at any level is to prevent the end result of the action of the affected gene.

Appendix 3A: Dependence of permeability of ions on various factors

Factors which have been observed to influence the variation of permeability of a specific substance in a specified membrane are composition, chemical reactions, and electric potentials. The equation of continuity for a species participating in a chemical reaction yields the influence of the chemical reaction on the fluxes of species participating in the chemical reaction. The influence of a chemical reaction on the fluxes of permeant or permeable species not participating in the chemical reaction should come from the coupling terms, such as the partial frictional coefficients, which link the flux of one species with the fluxes of other species in the system. In this appendix, the formal solutions of these basic questions of transport of species across charged membranes are briefly presented. For simplicity, the analysis is confined initially to charged ionic species, which are assumed to obey the Nernst-Planck kind of electrodiffusion equations.

Since, $\{d/dx\}[u(x)v(x)] = \{du/dx\}v(x) + u(x)\{dv/dx\}$, one obtains

$$\{d/dx\}[C_i(x)\exp(Z_ie\phi(x)/k_BT) =$$
$$e^{(Z_ie\phi/k_BT)}[\{dC_i/dx\} + (Z_ie/k_BT)C_i(x)\{d\phi/dx\}] \qquad (3A.1)$$

Thus, one has the result that the integral of the equation

$$\{dC_i/dx\} + (Z_ie/k_BT)C_i\{d\phi/dx\} + [J_i/D_i] = 0 \qquad (3A.2)$$

is

$$C_i(x) = C_i(o)e^{[(Z_i e/k_B T)\{\phi(o) - \phi(x)\}]} - I_i(x) \qquad (3A.3)$$

where

$$I_i(x) = e^{[-(Z_i e/k_B T)\Delta\phi(x)]} \int_o^x [J_i/D_i] e^{[+(Z_i e/k_B T)\Delta\phi(x)]} dx \qquad (3A.4)$$

In the preceding equations, $Z_i e$ is the charge of an ion of kind i, and $C_i(x)$ is its concentration at a location x in the inhomogeneous system. ϕ is the electric potential at x, and $\Delta\phi(x) = \phi(x) - \phi(o)$ is the difference in the electric potential in the system at two locations, o and x. J_i is the flux of ions along the x-direction, and D_i is its diffusion coefficient. k_B is the Boltzmann constant and T is temperature in absolute scale.

Under equilibrium conditions, when fluxes are zero, the integral of Equation 3A.2 becomes the Nernst equilibrium potential, given by Equation 3A.3 when the integral I_i has a value of zero. Equation 3A.2 is the Nernst-Planck equation, where the flux of ions of kind i is considered as a part arising from the concentration gradient, and another part arising from the electric potential gradient.

Thus, the result of Equation 3A.3 presents formally the way of computation of the difference in concentrations of a specified ionic species at two locations, in the inhomogeneous system, given that one has knowledge of the electric potential difference and the values of J_i and D_i in the stationary state. The Nernst equilibrium potential relates the concentration difference between two locations in the system of a specified ion i with valence charge number Z_i, with the electric potential difference at these two locations, $\Delta\phi(x)$. Equation 3A.3 denotes the influence of fluxes on these concentrations and the electric potential differences under stationary-state conditions. The two specified locations x and o could be on either side of a biological membrane, through which an ion of kind i can permeate.

The evaluation of the integral I_i requires knowledge of the positional dependence of the electric potential, flux, and the diffusion coefficient, which all appear inside the integral. If one makes the simple assumption that there occurs no chemical reaction in which species i participates in the system, then its flux J_i becomes independent of the position variable x in the system and is a constant. If one makes a second assumption that the diffusion coefficient can also be regarded as a constant, then the term $[J_i/D_i]$ can be taken outside the integral, thus leaving only the electric potential profile $\phi(x)$ inside the integrand required for the evaluation of the integral I_i.

If one now assumes that the electric potential can be regarded as linear, given by the simple expression

$$\phi(x) = \phi(o) + Ex \qquad (3A.5)$$

where E is the field, considered as a constant, one can evaluate the integral I_i easily. This approximation leads to the constant field equation of Goldman. Although these approximations are attractive, they do not yield the correct answer to the problem. The constant field assumption imposes the condition that there are no local-charge densities, which is unrealistic. In addition, since Equation 3A.2 is valid for every ion in the system undergoing transport, one has by summation over all ionic species the relation

$$\sum_{m=1}^{n} \{J_m/D_m\} + \{dA(x)/dx\} = [\varepsilon/4\pi k_B T]\{d^2\phi/dx^2\}\{d\phi/dx\}$$

where

$$A(x) = \sum_{m=1}^{n} C_m(x) \tag{3A.6}$$

In obtaining Equation 3A.6, Poisson's equation

$$\nabla \cdot \varepsilon(x)\, \nabla\phi = -(4\pi e) \sum_{m=1}^{n} Z_m C_m \tag{3A.7}$$

has been utilized. ε is the dielectric coefficient of the medium. Using the identity

$$\{d/dx\}[\{d\phi/dx\}^2] = 2\{d^2\phi/dx^2\}\{d\phi/dx\}$$

Equation 3A.6 can be integrated to obtain

$$\Delta x \sum_{m=1}^{n} \{J_m/D_m\} + [A(x_1) - A(x_2)]$$
$$= [\varepsilon/8\pi k_B T] < \{d\phi/dx\}^2|_{x_1} - \{d\phi/dx\}^2|_{x_2}\rangle \tag{3A.8}$$

where $A(x_1)$ is the sum of concentrations of all ionic species at location $x = x_1$. Since in almost all biological systems experimentally measured values (Hurlbut, 1971) indicate that $A(x_1)$ does not equal $A(x_2)$, it follows from Equation 3A.8 that the field at one location cannot equal the field at another location in the membrane system. This, therefore, invalidates the validity of the constant field assumption. When flux terms are absent in Equation 3A.8, this is known as *Maxwell's osmotic-balance equation*, which relates the osmotic pressure to the electric strain.

Historically, one attempted to describe the passive transport of a species across the membrane by the use of Fick's diffusion equation:

$$J_m = -D_m \nabla C_m \tag{3A.9a}$$
$$= -D_m\{\Delta C_m/h\} \tag{3A9.b}$$
$$= -P_m \Delta C_m \tag{3A.9c}$$

where ΔC_m is the difference in concentration of species m between the two sides of the membrane, h is the membrane thickness, D_m is the diffusion coefficient of species m, and J_m is its stationary-state flux, expressed in moles per unit area per unit time. Since one had no knowledge of the value of the diffusion coefficient in the membrane phase, and one could experimentally measure both the fluxes and the difference in concentrations of a specified species, one defined the permeability, P_m, of a species m in the membrane. Thus, permeability equals

$$P_m = -J_m/[\Delta C_m] \qquad (3A.10)$$

In order to compute the permeability of a specified species in a specified membrane, one needs the expression or the value of both the fluxes and the difference in the concentration of this species in a specified membrane. Values of the experimentally measured permeabilities of substances (namely, permeability of the same substance in different membranes as well as the values of permeabilities of different substances in the same membrane) began to appear in literature. Variations in the values of permeability of the same substance in the same membrane under different conditions were both experimentally observed and theoretically postulated in a somewhat ad hoc manner.

If one recalls that the diffusion coefficient of a particle in a diffusing medium is essentially a reflection of the resistance that molecules of this kind experience for transit in the medium, one has the relation between the frictional coefficient ζ_j of molecule j in the system under question related to its diffusion coefficient D_j by Einstein's relation

$$D_j = \{k_B T / \zeta_j\} \qquad (3A.11)$$

The frictional coefficient essentially arises from interparticle interactions, action at a distance, and is thus a measure of the interactions of all molecules present in the system to the motion of a molecule of a specified kind. Thus, one may write that

$$\zeta_j = \sum_i^n \zeta_{ij} C_i(x) \qquad (3A.12)$$

where the sum should be taken over all molecular species. Thus, in a inhomogeneous diffusion barrier, such as a biological membrane, the frictional coefficient is a function of composition and hence of position. This introduces additional complications in the evaluation of the integral I_i. However, this yields a simple explanation for the dependence of permeability of ionic species on concentration (Vaidhyanathan, 1964).

Thus, one has the plausible explanation for the observed dependence of permeability on the electric potential difference from Equations 3A.4 and 3A.10, and the dependence of permeability on the concentrations and the fluxes of different species present in the system. The subject of irreversible thermodynamics emphasizes the existence of coupling between fluxes of different kinds of species undergoing transport in a multicomponent fluid system across biological membranes under stationary-state conditions.

One can now proceed further to understand the dependence of permeability on chemical reactions. In the simple case, when ions of kind i participate in a chemical reaction in the membrane phase, evidently, under stationary-state conditions from equations of continuity one has

$$\{dC_i/dt\} + \nabla . J_i = \begin{array}{l} \text{Production or consumption} \\ \text{of species } i, \text{ by the} \\ \text{chemical reaction occurring} \end{array} \qquad (3A.13)$$

Thus, under steady-state conditions when the right-hand side of Equation 3A.13 is nonzero, the flux of species i, J_i, is a function of the position variable. Hence, the value of the integral I_i varies as a function of the kind of chemical reaction occurring. Again, therefore, the permeability varies (Vaidhyanathan, 1977) from one system to another system and from one condition to a different condition.

The variation of permeability of, say, a species i across a membrane when a chemical reaction in which species i does not participate is observed experimentally. The explanation for this observation is again obvious, since coupling terms dictate mutual interference of the fluxes of one species on another. A quantitative account for this behavior, although analytically somewhat involved because the equations are more lengthy, is not intractable.

Appendix 3B. Influence of fluxes on stationary-state concentration profiles and electric potential profiles

In the Debye–Huckel theory, the limiting expression for the chemical potential of an ionic species k in solutions is given by the expression:

$$\mu_k = \mu_k^o(T, p) + k_B T \ln C_k + Z_k e\phi - (e^2\kappa/\varepsilon) \qquad (3B.1)$$

where C_k is the concentration of ions of kind k in the solution, ε is the dielectric coefficient, Z_k is the signed valence charge number of ion k. e is the coulombic charge. $\mu_k^o(T, p)$ denotes the composition-independent part of the chemical potential. κ is the Debye–Huckel ionic atmosphere parameter of theory of strong electrolyte called the screening constant.

$$\kappa^2 = (4\pi e^2/\varepsilon k_B T) \sum_\sigma Z_\sigma^2 C_\sigma \qquad (3B.2)$$

In an inhomogeneous region, like the interfacial region adjacent to a biological membrane, the concentrations of ionic species as well as the dielectric coefficient are functions of the position variable x, so that the ion atmosphere parameter is also a function of position.

When the last term of Equation 3B.1 is negligible, one has the relation between the concentrations at two different locations of an ionic species in the system, related to the difference in the electric potential, by the relation

$$\{C_\sigma(x_1)/C_\sigma(x_2)\} = \exp\{Z_\sigma e[\phi(x_2) - \phi(x_1)]/k_B T\} \qquad (3B.3)$$

This is the Nernst equilibrium relation utilized to evaluate the electric potential difference by the measurement of the concentrations of a specified ionic species at two different locations. These two different locations can be on either side of a membrane. As stated earlier, the logarithmic dependence on the concentration of the chemical potential of an ion or solute in solution accounts for the entropy term contribution to free energy. There should be present terms accounting for the energetic contributions arising from interionic and interparticle interactions to free energy. The influence of the fluxes on the stationary-state concentration distribution is presented formally by Equations 3A.3, as a correction to the Nernst equilibrium distribution.

The approximations involved in the derivation of Equation 3B.1 regarding the potential of mean force are well known. If one assumes the validity of Equations 2.35 for a multicomponent solution, in an inhomogeneous region, the chemical potentials of ionic species σ and nonionic species j, may be written as

$$\mu_\sigma(x) = \mu_\sigma^o(T, p) + Z_\sigma e\phi(x) + k_B T \ln C_\sigma(x)$$

$$+ \sum_\eta C_\eta(x)H_{\sigma\eta} + \sum_j C_j(x)H_{\sigma j}$$

$$\mu_j(x) = \mu_j^o(T, p) + k_B T \ln C_j(x)$$

$$+ \sum_k C_k(x)H_{jk} + \sum_\sigma H_{j\sigma}C_\sigma(x) \qquad (3B.4)$$

The Greek subscripts are utilized to denote ionic species, and the Roman subscripts are utilized to denote nonionic species. We have consciously included the positional dependence of concentrations of species at locations, having in mind the extension of relations for applicability to inhomogeneous regions like interfacial regions near a membrane or a charged surface. The position variable x is defined normal to the y-z plane of the membrane surface.

If one neglects all other interactions other than coulombic for ion–ion interactions and further assumes that the molecular integrals, which are numbers, $H_{\sigma\eta}$, as position independent, one may express that $H_{\sigma\eta} = Z_\sigma Z_\eta e^2 H$, where H is a constant independent of charges. Similarly, inclusion of only the ion–dipole interactions enables one to express the interaction's contribution to the chemical potentials arising from nonpolar molecules to the chemical potential of ionic species σ as $H_{j\sigma} = Z_\sigma e H^\star$, where $H^\star$ is another constant.

Thus, the thermodynamic forces which are the gradients of the chemical potentials causing diffusive flows can be expressed as

$$\{d\mu_\sigma/dx\} = k_B T[d \ln C_\sigma/dx]$$

$$+ Z_\sigma e[\{dQ/dx\} + H^\star \sum_j \{dC_j/dx\}]$$

$$Q(x) = \phi(x) - (H\varepsilon/4\pi)\{d^2\phi/dx^2\}$$

$$\{d\mu_j/dx\} = k_B T[d \ln C_j/dx] + \sum_k H_{jk}\{dC_k dx\}$$

$$+ H^\star e \sum_\eta Z_\eta \{dC_\eta/dx\}$$

$$\{d^2\phi/dx^2\} = -(4\pi e/\varepsilon) \sum Z_\eta C_\eta(x) \tag{3B.5}$$

The last of Equations 3B.5 is the Poisson equation with the assumption that the dielectric coefficient may be regarded as constant.

The frictional coefficient of ionic species σ, ζ_σ, and of the nonionic species j, ζ_j, in the diffusion barrier can be written as (see Equation 3A.12):

$$\zeta_\sigma(x) = [k_B T/D_\sigma(x)]$$

$$= R_\sigma + C_\sigma \zeta_{\sigma\sigma} + \sum_{\eta \neq \sigma} \zeta_{\sigma\eta} C_\eta(x)$$

$$+ \sum_j C_j(x) \zeta_{\sigma j}$$

$$\zeta_j(x) = [k_B T/D_j(x)]$$

$$= R_j + C_j \zeta_{jj}$$

$$+ \sum_{k \neq j} C_k kj + \sum_\sigma C_\sigma \zeta_{\sigma j} \tag{3B.6}$$

where D_σ and D_j are the diffusion coefficients of species j in the diffusion barrier. The terms R_σ and R_j are the contributions to ζ_σ and ζ_j arising from the immobile molecule of the membrane phase. The partial frictional coefficient ζ_{jk} represents the contribution of kth kind of molecules to the frictional coefficient of a molecule of kind j, through intermolecular interactions.

Under stationary-state conditions when there are no chemical reactions, the expressions for the thermodynamic forces, namely, the gradients of chemical potentials, satisfying the Gibbs-Duhem relation for the stationary states, (Equation 2.24), give a required second set of equations:

$$\{d\mu_\sigma/dx\} = -[J_\sigma\zeta_\sigma(x)/C_\sigma(x)] + \sum_{\eta\neq\sigma} J_\eta\zeta_{\sigma\eta}$$

$$+ \sum_j J_j\zeta_{\sigma j}$$

$$\{d\mu_j/dx\} = -[J_j\zeta_j(x)/C_j(x)]$$

$$+ \sum_\sigma J_\sigma\zeta_{\sigma j} + \sum_k J_k\zeta_{jk} \tag{3B.7}$$

where J_j and J_σ are the stationary-state fluxes of molecules of kind j and σ, respectively.

Specifically, for a system containing n kinds of ionic species and one kind of uncharged species, say, solvent molecules j, one has

$$\{dC_\sigma/dx\} + [Z_\sigma e/k_B T]C_\sigma(x)[\{dQ/dx\} + H^\star\{dC_j/dx\}]$$

$$+ R_\sigma^\star - S_{\sigma\sigma}C_\sigma + \sum_{\eta\neq\sigma} S_{\sigma\eta}C_\eta + S_{\sigma j}C_j = 0$$

$$\{dC_j/dx\} + (H_{jj}/k_B T)C_j\{dC_j/dx\} + R_j^\star - S_{jj}C_j + \sum_\sigma S_{j\sigma}C_\sigma$$

$$= [H^\star\varepsilon/4\pi k_B T]C_j(x)\{d^3\phi/dx^3\}$$

$$R^\star = \{J_\sigma R_\sigma/k_B T\}; \qquad R_j^\star = \{J_j R_j^\star/k_B T\};$$

$$S_{\sigma\eta} = \{J_\sigma\zeta_{\sigma\eta}/k_B T\}; \qquad S_{j\sigma} = \{J_j\zeta_{\sigma j}/k_B T\};$$

$$S_{\sigma\sigma} = \left[\sum_{\eta\neq\sigma} S_{\eta\sigma} + S_{j\sigma}\right];$$

$$S_{jj} = \sum_\sigma S_{\sigma j} \tag{3B.8}$$

In the absence of chemical reactions occurring in the diffusion barrier, $S_{\sigma\eta}$ and $S_{\sigma j}$ are constants independent of x. The stationary-state electric potential profile and the concentration profiles are obtained by solution of the differential equations of Equations 3B.8. When one neglects the composition and position dependence of the frictional coefficients and ignores $H^\star$ terms, one has from the first of Equations of 3B.8

$$\{dC_\sigma/dx\} + [Z_\sigma e/k_B T]C_\sigma(x)\{dQ/dx\} + \{J_\sigma/D_\sigma\} = 0 \qquad (3B.9)$$

When one neglects in addition the H term, that is, setting the activity coefficient of ions of kind equal to unity, one has the familiar Nernst–Planck equation for electrodiffusion. It is possible to obtain at least numerical solutions of the preceding set of coupled nonlinear differential equations, provided the values of the partial frictional coefficients and the stationary-state fluxes are known from experimental measurements.

REFERENCES

BERENDSON, H.J.C., in *Theoretical and Experimental Biophysics*, ed. A. Cole, New York: Marcel Dekker, Inc., 1967.

DEL CASTILLO, L.F., and E.A. MASON, *Biophysical Chemistry*, vol. 12, (1980), p. 223.

FORSLIND, E. *Quant. Rev. Biophysics*, vol. 4, (1971), p. 352.

FRANK, H.S., *J. Chem. Phys.*, vol. 23, (1955), p. 2023.

FROHLICH, H., *Biological Coherence and Response to External Stimuli*. Berlin: Springer-Verlag, 1988.

HORNE, R.A., *Water and Aqueous Solutions*. New York: Wiley Interscience, 1972.

HURLBUT, W.P., in *Membranes and Ion Transport*, vol. 2, ed. E. Bittar. New York: Academic Press, 1971.

KATCHALSKY, A., and P.F. CURRAN, *Nonequilibrium Thermodynamics in Biophysics*, Chapter 10. Cambridge, MA: Harvard University Press, 1967.

MUNTWYLER, E., *Water and Electrolyte Metabolism and Acid-Base Balance*. St. Louis: The C. V. Mosby Company, 1968.

RAHN, H., and O. PRAKASH, *Acid-Base Regulation and Body Temperature*. Boston: Martinus Nijhoff, Publishers, 1985.

REEVE, E.B., and L. KULHANEK, in *Physical Basis of Circulatory Transport*, eds. Reeve and Guyton. Philadelphia: W. B. Saunders Co., 1967.

VAIDHYANATHAN, V.S., *J. Theor. Biology*, vol. 7, (1964), p. 334.

VAIDHYANATHAN, V.S., *J. Theor. Biology*, vol. 10, (1966), p. 159.

VAIDHYANATHAN, V.S., in *Electrical Phenomena at the Biological Membrane Level*, ed. E. ROUX, p. 113. Amsterdam: Elsevier Scientific Publishing Company, 1977.

WARNER, J. *Theor. Biology*, vol. 1, (1961), p. 514.

WARNER, J. *Theor. Biol*, vol. 6, (1964), p. 118.

chapter 4

Aspects of System Behavior in Stationary States

Living systems obtain energy from external sources and must control energy expenditure internally to exhibit their organized behavior. Living systems should control their internal environment efficiently. Homeostasis necessarily implies time–independent behavior. Therefore, the behavior of systems under stationary–state conditions becomes relevant. One finds almost continual fluctuations in the environment. Thus, biological systems are subject to disturbances, and the system undergoes changes based on the variations in system parameters.

The laws of thermodynamics and the equations of continuity and conservation impose conditions which tend to preserve the system in its old state, in a new time–independent stationary state, or in an oscillatory state depending on external and dissipative forces. With each disturbance, there can exist transient changes. Following one such perturbation, the behavior of the system will be that of the disturbed system and this new state can be quite different from the initial state of the system. The disturbances can change the form of the system equations and thus affect all aspects of the system response and its behavior. In the respiratory system, for example, an increase in muscular work will increase the rate of production of carbon dioxide and cause an increase in the ventilatory rate.

Fluid dynamics concerns itself with the study of the motion of fluids. The phenomena considered in fluid dynamics are macroscopic, and a fluid can be regarded as a continuous medium. Thus, any small element of volume in the system is assumed to be sufficiently large enough to contain a large number of molecules. The behavior of the system can be analyzed with the use of constraints imposed by the equations of continuity, the equations of conservation, the laws of thermodynamics and chemical kinetics, as well as the feedback behavior existing in the system due to prevailing interconnections between various component compartments of the system. The mathematical description of the state of the moving fluid is effected by means of functions which give the distribution of the fluid velocity and of any two of the thermodynamic quantities pertaining to the fluid, for example, the density and pressure.

The discipline of thermodynamics applied to nonequilibrium situations prescribes relationships among a collection of interacting forces and fluxes without regard to the mechanisms of these interactions, or to the structure of the system. With the imposition of some external constraints, such as the maintenance of concentration differences across membranes, or the electrical potential differences, the forced evolution of the system ultimately leads to a stationary state. So long as systems, including biological systems, consist of molecules and ions with accompanied appropriate interparticle interactions, the system when perturbed freely evolves in such a manner that it tends to reach a unique time–independent stationary state called the thermodynamic equilibrium state. The response of the system arising from interparticle interactions tends to oppose the external force and bring the system into a nonequilibrium stationary state, when the influence of the external force balances the response of the system.

Whenever a force acts on an object, it undergoes no change except compression, and remains in equilibrium if the vector sum of all forces acting on the body is zero. Equilibrium in this situation implies that the object either remains at rest or continues to move with a constant vector velocity. If the object is in rotational equilibrium, this implies that the object does not rotate, or continues to rotate at a constant rate. When the input and output flow rates of a system have been identified and the applicable conservation relation is known in the steady state, these flow rates will either equal or possess a fixed relation to each other. In many processes, the input and output variables may not be known and the application of any conservation law may not be obvious. The behavior of a system can be analyzed with the use of conditions of constraints imposed by the equations of continuity, the laws of thermodynamics and chemical kinetics, as well as the feedback behavior existing in the system due to prevailing interconnections between various subsystems of the system.

STABILITY CONSIDERATIONS _____

The term *stability* has a number of meanings and is used in different ways. Qualitatively, stability frequently connotes a certain constancy and a lack of change. More technically, stability may mean the ability to withstand disturbances without disruption or permanent loss of normal function. An important principle, called the Le Chatelier-Braun principle, states that whenever a stress is placed on a system, the system is constrained to move in such a direction so as to nullify the effect of stress. It may be recognized that biological systems, including their repair mechanisms, most likely function in this manner. An extension of this principle to stationary states has been formulated by Prigogine and De Groot.

In thermodynamic systems, three kinds of equilibrium are evident. A thermodynamic system is in stable equilibrium if the condition

$$\Delta E > 0 \tag{4.1}$$

holds for all conceivable variations in state. The quantity ΔE is the change in energy of the system when it passes from a given equilibrium state to a neighboring state, under the constraints of constant volume, entropy, and number of moles of the components. A system is in neutral equilibrium if the condition

$$\Delta E = 0 \tag{4.2}$$

holds for some variations and the condition in Equation 4.1 holds for other variations. A thermodynamic system is in unstable equilibrium if the condition

$$\Delta E < 0 \tag{4.3}$$

holds for any conceivable variation in state. In order that Equation 4.1 is satisfied for the system, for a continuous change, the first nonvanishing variation, δE, (of the Taylor series expansion of E of the system around the equilibrium state) must be positive definite. For stable equilibrium, the second variation, $\delta^2 E$ should also be greater than zero. For unstable equilibrium, this second variation should be negative definite (Kirkwood and Oppenheim, 1961).

The question of stability or instability becomes an important aspect in systems with feedback, since the system may become unstable and disruptive even though it is stable in the absence of the feedback. The homeostatic processes are normally stable, having acquired this property during normal evolutionary pressures. The concept of potential instability for physiological systems as well as the boundary between normal stable behavior and instability is of scientific as well as clinical interest. With a given definition of stability, it has been possible to develop criteria for the existence of stability, expressed as quantitative mathematical relations between the various system parameters.

A system is said to be *asymptotically stable* if following a disturbance, the system returns asymptotically to its previous steady state, or if the disturbance is prolonged, to a new steady state. An *absolutely stable* system is one that remains stable for all values of its parameters. A system is said to be *conditionally stable* if it is stable for both high and low values of loop gain, but it is unstable for intermediate values of gain. *Orbital stability* refers to the ability of an oscillating system to maintain its frequency and amplitude.

The preceding definitions have been taken from the mathematical theory of stability, which is detailed later in this book. The term *global stability* refers to the ability of a system to remain stable for all possible normal operating conditions. Physiological regulators must obviously exhibit global stability in that they are stable under all normal operating conditions. They need not be absolutely stable.

Stability is a property of the entire system, its interconnections, and the magnitude of all its parameters, and is not affected by the disturbances imposed by the environment. Homeostatic processes are normally stable except for pathological cases. The mere complexity of homeostatic systems raises the basic question of how stability is in fact achieved. Until this question is answered, at least partially, one cannot feel secure that our understanding of these systems is in any sense adequate. Stability is usually discussed in systems theory on the basis of a model of a prototype system, and the conclusions are only as valid as the model. Stability becomes one measure of the competence of the model. If both the model and prototype can be thrown into instability by the same parametric changes, there is reason to have greater confidence in the mathematical representation.

EQUATION OF CONTINUITY

Since many of the phenomena considered in fluid dynamics are macroscopic, a fluid system can be regarded as a continuous medium. Thus, any small element of volume in the system is assumed to be sufficiently large to contain a large number of molecules. The mathematical description of the state of a moving fluid is effected by means of functions which give the distribution of the fluid velocity, $V = V(x, y, z, t)$ and of any two thermodynamic quantities pertaining to the fluid, for example, the density ρ and pressure p, which are also functions of the three Cartesian coordinates and time. The equation of state describes the system in terms of two independent variables, say pressure and temperature, while the third, namely, volume V is automatically determined. The thermodynamic properties of a one-component fluid are determined by the values of any two of them, together with the equation of state. Thus, with the three Cartesian components of the velocity v, the pressure p, and the density ρ, the state of the moving fluid is completely determined.

The matter flux J_k of a substance k is defined as the amount of substance transported across a plane per unit area per unit time. If the substance k does not participate in a chemical reaction, it cannot appear or disappear, and its mass is conserved. This conservation leads to the validity of the *equation of continuity*. The equation of continuity states that the rate of decrease of the amount of a substance in a certain region equals the rate at which it leaves the region as flow through the surface surrounding the region. For one-dimensional flow, it is written as

$$\{\partial C_k / \partial t\} = - \{\partial J_k / \partial x\} \tag{4.4}$$

where C_k is the number density of species k, which is a function of both position variable and time. Flux is a vector with three Cartesian components. Thus, for three-dimensional transport, one writes the equation of continuity as

$$\{\partial C_k / \partial t\} + \nabla \cdot J_k = 0 \tag{4.5a}$$

$$\nabla. = \{\partial / \partial x\} + \{\partial / \partial y\} + \{\partial / \partial z\} \tag{4.5b}$$

The vector operator, $\nabla.$, is called the divergence. Evidently, if there occurs a chemical reaction in the system, in which species k participates, then the right-hand side of Equation 4.5a cannot equal zero. Instead, a term accounting for the production or consumption of k should be present. Under stationary-state conditions, it is evident that the time derivative of density vanishes, and thus the divergence of flux term vanishes in Equation 4.4. Hence, the matter flux J_k is a constant independent of the position variable, as expected in the steady state, if there is no chemical reaction in which species k participates.

Assume that in a system a chemical reaction of the kind

$$A + B \underset{k_2}{\overset{k_1}{\rightleftharpoons}} C \tag{4.6}$$

where species A and B are the reactants and C is a product occurs in the volume elements of the system. If a, b, and c denote the local concentrations of the three species, one has from standard reaction kinetics that the rate of chemical reaction J_r is given by the relation

$$J_r(x) = k_1 a(x) b(x) - k_2 c(x) \tag{4.7}$$

where k_1 and k_2 are the rate constants for the forward and reverse reactions. The equilibrium constant K is defined as the ratio $[k_1 / k_2]$. x is a position variable, along one of the three Cartesian axes.

Under equilibrium conditions, J_r equals zero. In general, under non-equilibrium conditions, J_r is a function of both time and position. Under stationary-state conditions, the time dependence of J_r vanishes. Thus, under stationary-state conditions, when a chemical reaction of the kind in Equation

4.6 occurs in the system, the fluxes of the three chemical species as well as the concentration profiles are functions of the position variable. In the inhomogeneous system, this could be the diffusion barrier. The partial equations of continuity, (Equations 4.5), now become,

$$\{\partial a / \partial t\} + \nabla \cdot J_a = -J_r \tag{4.8a}$$

$$\{\partial b / \partial t\} + \nabla \cdot J_b = -J_r \tag{4.8b}$$

$$\{\partial c / \partial t\} + \nabla \cdot J_c = +J_r \tag{4.8c}$$

The influence of chemical reactions on fluxes of species across membranes can best be analyzed with the use of these equations of continuity. Equations 4.8 are needed for the analysis of the carrier transport model in a membrane, when reactions of the kind in Equation 4.6 are presumed to occur. B is the carrier and C is the carrier-substrate complex, with A being the substrate transported across the diffusion barrier. It is usually assumed that the carrier and the complex are confined to the diffusion barrier. This implies that the fluxes of B and C at the boundaries of the membrane phase assume null values. Addition of Equations 4.8a and 4.8c for stationary-state conditions when time derivatives vanish yield

$$\nabla \cdot \{J_a(x) + J_c(x)\} = 0 \tag{4.9}$$

Therefore, $[J_a(x) + J_c(x)]$ is a constant, independent of x. Assuming that fluxes are adequately described by Fick's law for diffusional flow,

$$J_j = -D_j\{dC_j/dx\} \tag{4.10}$$

where D_j is the diffusion coefficient of species j in the system and C_j is its concentration, x is a position variable defined along the x-axis, along which transport occurs, normal to the plane of the diffusion (membrane) barrier, and J_j is its matter flux. Assuming that the positional variation of the diffusion coefficients if any in the system can be neglected, one has

$$D_a\{da/dx\} + D_c\{dc/dx\} = N, \quad \text{a constant} \tag{4.11a}$$

With one more integration, one has

$$D_a \, a(x) + D_c \, c(x) = M + Nx \tag{4.11b}$$

Similarly, one has for the species B and C,

$$D_b \{db/dx\} + D_c \{dc/dx\} = \text{another constant}$$

Assuming that the diffusion coefficients of B and C equal each other, one has

$$\{db/dx\} + \{dc/dx\} = \text{another constant} \tag{4.12}$$

If the reaction represented by Equation 4.6 is presumed to occur in a membrane phase, and species B represents the carrier molecules, species A represents the permeant species, and species C denotes the carrier-substrate complex, one has a quantitative manner of analyzing the carrier model. One assumes that the carrier molecules B and the carrier-substrate complex C are confined to the membrane phase. Thereby, one obtains the result,

$$\{b(x) + c(x)\} = C_T - (J_b + J_c)x \tag{4.13}$$

where x is the position variable normal to the plane of the membrane and along the axis of transport. C_T is the total concentration of the carrier molecule. For extremely inhomogeneous regions like the membranes of biological systems, the assumption of position-independent diffusion coefficients is unlikely to be valid.

APPLICATION TO CARRIER TRANSPORT MODEL

It is desirable to be able to compute the influence of chemical reactions on fluxes across membranes under stationary state. If one has knowledge of the concentration profiles of A, B, and C as a function of position variables, in an inhomogeneous system, in which the reaction specified in Equation 4.6 occurs, evidently, the reaction rate profile $J_r(x)$ can be computed with the use of Equation 4.6. The concentration profiles $C_i(x)$ are a priori unknown.

If one has knowledge of the reaction rate profile, a simple integration will yield the profiles of fluxes of the three species in the reaction medium, apart from the knowledge of the integration constants. Evidently these constants may be identified as the fluxes of the species, across the same diffusion barrier, for the same difference in concentrations existing across the diffusion barrier, when there is no chemical reaction. They will also equal the value of the stationary state fluxes of respective species, if the reaction in question is at equilibrium at the boundaries of the diffusion barrier.

When one has knowledge of the flux profiles of the three species in the diffusion barrier, provided that one can assume a simple relation like Fick's law, then with an additional integration one can compute the concentration profiles of the three species participating in the chemical reaction in the diffusion barrier (Equation 4.6). Thus, to evaluate the influence of chemical reactions on stationary-state fluxes, one has the vicious cycle of requiring either the knowledge of the concentration profiles, or the reaction rate profiles from the start and compute the rest. This procedure is applicable to all chemical reactions known to occur in the diffusion barrier. Understanding the influence of chemical reactions on fluxes is essential to understand basic transport processes occurring in biology, such as facilitated transport or active transport.

For biological transport across membranes in the carrier model, one presumes that both the carrier and the carrier-substrate complex are confined to the membrane phase. Thus, the reaction is also confined to the membrane phase. Physically, one assumes that the substrate molecules A undergo diffusion due to the existence of concentration differences across the membrane. In addition, they combine with the carrier molecules B present in the membrane to form a complex C. Both A and C molecules transport across the membrane from one side to the other. At the other boundary it is presumed that the complex C decomposes back to A and B. The molecules A obtained by dissociation go to the other side, while the B molecules produced are retained in the membrane phase. Since more B are formed at the second boundary, and B molecules are consumed at the first boundary to form C, there develops a natural gradient in concentration of B in an opposite direction to the concentration gradient of A. Thus, B diffuses back to the first interface to play its role again.

A simple mathematical analysis of this model has been proposed by Blumenthal and Katchalsky (1969). From Equations 4.9 and 4.10 under stationary states, one has the simple relations

$$
\begin{aligned}
J_r(x) = \{dJ_C/dx\} &= -D_C\{d^2c/dx^2\} \\
&= -\{dJ_A/dx\} = D_A\{d^2a/dx^2\} \\
&= -\{dJ_B/dx\} = D_B\{d^2b/dx^2\}
\end{aligned}
\tag{4.14}
$$

From Equation 4.7, one has in addition

$$
\begin{aligned}
\{dJ_r/dx\} &= k_1[a(x)\{db/dx\} + \{da/dx\}b(x)] - k_2\{dc/dx\} \\
\{d^2J_r/dx^2\} &= k_1[a\{d^2b/dx^2\} + b\{d^2a/dx^2\}] \\
&\quad - k_2\{d^2c/dx^2\} + 2k_1\{da/dx\}\{db/dx\}
\end{aligned}
\tag{4.15}
$$

The last term in the right-hand side of Equation 4.15 is a nonlinear term. Neglecting this term, and replacing the second derivatives of concentration terms in the second of Equation 4.15 with the results obtainable from Equations 4.14, one has the simpler differential equation

$$
\begin{aligned}
\{d^2J_r/dx^2\} &= Z^2(x)J_r(x) \\
Z^2(x) &= k_1[\{aD_B + bD_A\}/\{D_AD_B\}] + \langle k_2/D_C \rangle
\end{aligned}
\tag{4.16}
$$

The first of Equation 4.16 is a simple second-order differential equation for the reaction rate profile J_r, provided the term Z is a constant independent of x. Unfortunately this is not the case, since the definition of Z involves the concentration profiles, a, b, and c, which are evidently functions of position. Thus, neglect of the nonlinear term, namely, the last term of Equation 4.15, is not sufficient to solve the problem of obtaining the reaction rate profile. Blumenthal and Katchalsky utilized the additional approximation that the concentration terms a, b, and c of Equation 4.16 can be replaced by the equilibrium concentrations of the reaction of Equation 4.6, thus forcing the term Z to be independent of the position variable. When such approximation is valid, the solution of first of Equation 4.16 is simple and the results are presented by Blumenthal and Katchalsky. They obtain a reaction rate profile which is symmetric about the midpoint of the diffusion barrier. In obtaining their results, they utilize the additional boundary condition that

$$\int J_r(x) \, dx = 0 \tag{4.17}$$

For dimensional consistencies, the integral of Equation 4.17 is in reality a volume integral. The errors incurred in the neglect of the nonlinear term as well as approximating concentrations in the definition of the relaxation length Z by equilibrium concentrations are not known. Blumenthal and Katchalsky obtain the following result for the reaction rate profile, as a solution of Equation 4.16,

$$J_r(x) = J_r^o\{\cosh\langle x/Z \rangle - \coth\langle x/2Z \rangle \sinh(x/Z)\} \tag{4.18}$$

where J_r^o is the rate of chemical reaction at $x = 0$.

Twice integration of the reaction rate profile, Equation 4.18 yielding the concentration profiles, $a(x)$, $b(x)$, and $c(x)$, when substituted in Equation 4.7 does not yield back an expression for the reaction rate profile as a consistent result, which should agree with the result of Equation 4.17. Therefore, the neglect of the nonlinear term, as well as regarding Z as independent of x, is not justified. Without these approximations, one obtains a more complicated differential equation for the reaction rate profile (Vaidhyanathan, 1971):

$$J_r(x) = \{d^2G/dx^2\}$$
$$= N^2G + MG^2 + S \times G + px^2 + qx + J_r(o)$$

where

$$N^2 = [k_1\{a(o)D_B + b(o)D_A\}/\{d_AD_B\}] + (k_2/D_C)$$
$$M = \{k_1(D_AD_B)\}$$
$$S = M(Q_A + Q_B)$$
$$p = MQ_AQ_B$$
$$q = M\{a(o)D_AQ_B + b(o)D_BQ_A\} + \langle k_2Q_C/D_C \rangle$$
$$J_r(o) = \text{Reaction rate at location } x = 0, \text{ i.e., one of the}$$
$$\quad\quad \text{boundaries of the diffusion phase}$$
$$Q_A = -[I + J_A]$$
$$Q_B = -[I + J_B]$$
$$Q_C = J_C - I$$
$$\{dI(x)/dx\} = J_r(x) \tag{4.19}$$

Equations 4.19 are valid, so long as fluxes can be described by Fick's law in the membrane phase, and the diffusion coefficients of species in the membrane phase can be regarded as position independent. $J_r(o)$, which represents the reaction rate at location $x = 0$, namely, one boundary of the membrane is nonzero. One must solve equations of the kind in Equations 4.19 to obtain a quantitative estimate of the influence of chemical reactions on fluxes across a diffusion barrier under stationary-state conditions. The problem of obtaining the reaction rate profile in the membrane phase, even when a simple reaction of the kind in Equation 4.6 occurs, is necessarily complicated.

THERMODYNAMICS OF IRREVERSIBLE PROCESSES

The statement that $dS > \{dQ/T\}$ is valid for only closed systems and cannot be utilized to compute the entropy increase occasioned by an irreversible process. Visualization of entropy as a fluid capable of flowing from one part of space to another, and the concept that entropy can be destroyed and created leads one to the entropy-balance equation. The entropy production is the amount of entropy created per unit time. Only for an isolated system this equals $\{dS/dt\}$. The entropy source density σ is the entropy production per unit volume. The entropy flux density J_s is a vector that coincides with the direction of entropy flow and has magnitude equal to the entropy crossing per unit area perpendicular to the direction of flow per unit time. Both σ and J_s will in general be functions of position and time. The entropy flux vector J_s, the entropy production per unit volume σ, and the specific entropy s are related to each other by the following equation.

[rate of increase of entropy   [outward flux of entropy
  inside a volume element dV] $+$ from dV]

$$= \{\text{entropy production inside } dV\},$$

$$\{\partial \rho s / \partial t\} + \text{div} J_s \, dV = \sigma \, dV \qquad (4.20)$$

This equation is called the *entropy–balance equation*.

In nonequilibrium states, forces cause fluxes. The experimental basis for the thermodynamics of irreversible processes is provided by many phenomenological laws establishing the proportionality between the rate of some process and the forces responsible for its occurrence. The relation of heat flow to the temperature gradient known as the Fourier law is an example. The relation between the current strength with the difference in electric potential is given by Ohm's law. In addition, the possibility of one process affecting or interfering with another process exists, when there is more than one kind of force. Under stationary state, when one can presume that forces are proportional to fluxes, one has the general relation

$$J_k = L_{kk} X_k \qquad (4.21)$$

where X_k is the force acting on species k, which is normally the gradient of a scalar potential, and J_k is the flux of species k. L_{kk} is a phenomenological coefficient not dependent on either J_k or X_k. This cause-effect relation between force and flux and identification of X_k as a thermodynamic force, as well as establishment of their relation to entropy production enables one to formulate the thermodynamics of irreversible processes.

When n processes occur at the same time, the associated fluxes J_i are linearly dependent on the conjugate forces X_i,

$$J_i = \sum_{j=1}^{n} L_{ij}X_j; \qquad i = 1, 2, \ldots, n \qquad (4.22)$$

where

$$L_{ij} = \{\partial J_i / \partial X_j\}_{X_k} : k \neq j \qquad (4.23)$$

are partial derivatives at the point $X_1 = X_2 = \cdots = X_n = 0$. The assumption that Equation 4.23 represents the generalized phenomenological relations is a basic postulate of the theory. The terms appearing in the right-hand side of Equation 4.23 are only the first-order contributions in a Taylor expansion of J_i as a function of the forces X_j. Thus, Equation 4.23 is valid when the nonequilibrium stationary state is very close to equilibrium, or when changes involving forces are very small compared to unity.

The rate of entropy production for a process involving n fluxes is given by the expression

$$\sigma = \sum_{i=1}^{n} J_i X_i \qquad (4.24)$$

This is a bilinear form in forces and fluxes. Substitution of Equation 4.23 in Equation 4.24 leads to the expression for entropy production as a quadratic form in the forces alone:

$$\sigma = \sum_{i}^{n} \sum_{j}^{n} L_{ij}X_i X_j \qquad (4.25)$$

Thus, the linear thermodynamics of irreversible processes is developed based on the assumed validity of three postulates:

1. The entropy production can be computed from Gibb's relation.
2. The entropy production, written as the product of forces and fluxes, is nonnegative.
3. Each flux entering in the expression for the entropy production is a linear combination of forces.

Onsager proposed a theorem that the coefficients L_{ij} of the phenomenological equations are symmetrical.

$$L_{ij} = L_{ji}$$
$$\{\partial J_i/\partial X_j\}_{X_{k \neq j}} = \{\partial J_j/\partial X_i\}_{X_{j \neq k}} \qquad i, j = 1, 2, \ldots n$$

(4.26)

Thus, Onsager's reciprocal relations state that the increase in flux of i, J_i caused by unit increase in the force X_j, while the remaining forces are held fixed, is equal to the increase in the flux J_j due to unit increase in the force X_i. The resemblance between Onsager's relation and Maxwell's equations in Chapter 2 is evident. Onsager's relations may also be regarded as a statement of the impossibility of creation of energy or perpetual motion. Long before Onsager's postulation, the reciprocal relations were suggested by observations such as the Peltier and Seebeck effects, and by experiments on the flow of heat in anisotropic crystals.

The theory of nonequilibrium thermodynamics, as developed by Onsager, is based on the postulate that the entropy S of an isolated system is given by

$$S = (1/2T) \sum_i \sum_k g_{ik} a_i a_k$$

(4.27)

where a_i and a_k are the thermodynamic variables (system parameters) needed to describe the state of the system. The constant coefficients obey the symmetric relation

$$g_{ik} = g_{ki}$$

(4.28)

The time derivative of entropy can be written as

$$\sigma = T\{dS/dt\}$$
$$= (1/2)\{d/dt\} \sum_i \sum_k g_{ik} a_i a_k$$
$$= \sum_i \sum_k g_{ik}\{da_i/dt\} a_k$$

(4.29)

Defining flux J_k as the time derivative of the thermodynamic variable a_i and the conjugate force X_i by the relation

$$X_i = T\{\partial S/\partial a_i\}$$
$$= \sum_k g_{ik} a_k$$
$$J_i = \{da_i/dt\}$$

(4.30)

One writes the dissipation function as

$$\sigma = \sum_i J_i X_i$$

(4.31)

These are the definitions of generalized forces and fluxes in terms of entropy and thermodynamic variables of the system. Though the choice of forces X and fluxes J of Equation 4.30 are not necessarily unique, they must satisfy certain requirements:

1. The forces should be so chosen that they vanish at equilibrium state.
2. The fluxes can be expressed as a vector-valued function of the thermodynamic forces X, and of the state variables of corresponding thermodynamic equilibrium state.
3. These fluxes should also vanish with vanishing forces at equilibrium.
4. The fluxes and forces are such that the product of conjugate forces and fluxes is positive definite, which is a requirement that follows from the second law of thermodynamics. These four requirements are not necessarily restricted to the linear region.

Essentially, three major postulates form the basis of the derivation of the reciprocal relations by Onsager.

1. For a system in which irreversible processes are occurring, all thermodynamic functions exist for each element of the system. These thermodynamic functions for nonequilibrium systems are the same functions of the local state variables as the corresponding equilibrium thermodynamic quantities. This is called the *assumption of local equilibrium*. The local quasi-equilibrium hypothesis is a precondition for the applicability of the phenomenological description both in the linear and nonlinear regions. In the nonlinear region there is no sufficiently adequate general scheme based on this hypothesis.

2. The second postulate is that the fluxes are linear homogeneous functions of the forces. Thus, one can write that any flux J_i is related to all the forces present by the relation

$$J_i = \sum_k L_{ik} X_k \qquad (4.32)$$

where the phenomenological coefficients L_{ik} are independent of the forces.

3. The third postulate is that the matrix of phenomenological coefficients is symmetric.

Onsager has considered the general problem of entropy production and symmetry properties of the phenomenological coefficients in the linear relations. This analysis is based on consideration of the regression of fluctuations in systems away from equilibrium. The symmetric nature of the phenomenological coefficients arises from the principle of microscopic reversibility.

According to Boltzmann, the seeming irreversibility of natural changes is attributable to the fact that the system will proceed to states of greater probability when it happens to be in states of lesser probabilities. The state of thermodynamic equilibrium has by far the greatest probability of all the states compatible with the condition of equilibrium. Fluctuations of appreciable size are very rare events.

$$W = W_o \exp\{\Delta S / k_B\} \qquad (4.33)$$

where $\Delta S = S - S_o$ represents the excess entropy S over the equilibrium state value S_o. It is necessarily negative or zero, since the entropy is maximum in the equilibrium state. The probability W decreases very sharply with increase in S, since the Boltzmann constant k_B is an extremely small quantity, of the order of 10^{-16}. The state is defined by special values of the set of parameters, a_1, a_2, ..., a_n, which is called the Onsager coordinates. It is required that instead of representing local quantities, these Onsager coordinates designate the deviations of these properties from their equilibrium values.

Onsager defines the functions

$$F_i = \{\partial S / \partial a_i\} = - \sum_j g_{ij} a_j \qquad (4.34)$$

where F_i are called the Onsager forces. Both Onsager forces and Onsager coordinates a_i vanish at equilibrium. The rate of entropy production $\{dS/dt\} = \dot{S}$ can be expressed in a bilinear form in the forces and the time derivatives of the Onsager coordinates, a_i.

$$\dot{S} = \dot{a}_i F_i \qquad (4.35)$$

Onsager's proof of the validity of reciprocal relations based on three considerations.

1. *Principle of detailed balancing*: When two or more independent chemical reactions take place in a system, chemists are accustomed to regard each reaction as balanced separately at equilibrium. This is called detailed balancing of chemical reactions and thus constitutes an additional hypothesis not derivable from thermodynamics.

2. *Principle of microscopic reversibility*: Under equilibrium conditions, any molecular process and its reverse take place on the average with the same frequency.

3. *Regression of fluctuations*: A deviation from equilibrium, which occurs spontaneously as a result of a fluctuation, decays on the average in the same manner as a deviation that has been artificially induced by outside interference.

It is appropriate to emphasize that a phenomenological characterization of total observed phenomena, including the statement that processes occurring in biology are interrelated and hence are interdependent and coupled, is accomplished by the use of concepts of irreversible thermodynamics to biological systems. Just like in equilibrium thermodynamics, it cannot inherently present any insight into the mechanism of processes occurring in the system. They present only the overall constraints, which may be interpreted as controlling mechanisms in biology.

Onsager's formalism of irreversible thermodynamics relates the thermodynamic flows to their conjugate forces. Sufficiently close to equilibrium, the constitutive relations between forces and fluxes become linear, and the matrix of the phenomenological coefficients is symmetric in most cases. Admitting that living systems are extremely complicated, heterogeneous, and nonlinear systems, subtle interplays that exist between the energetic rate processes of transport, chemical reactions, and free energy dissipation should be recognized and identified.

There are a number of important problems which appear to be, in principle, inaccessible to the phenomenological approach. There is the problem of defining the limits of applicability of the thermodynamic description. There is also the problem of an ambiguity in the definition of a nonequilibrium thermodynamic entropy, and the even greater problem of whether or not the concept of entropy exists at all for systems far from equilibrium. In order to formulate a general phenomenological scheme of irreversible thermodynamics far from equilibrium, it is necessary first to formulate in a sufficiently general form the hypothesis of probability distribution which would specify and generalize the local equilibrium hypothesis. A theory of nonequilibrium states has been proposed by Meixner (1963) based on a general theory of passive systems. A truly nonlinear field theory has been developed by Truesdell (1969). Many attempts have been made to extend the treatment of coupled, nonlinear, time-independent thermodynamic processes by employing graphical notations, emphasizing the topological relations of the system. This is called *network thermodynamics*. In this approach, a graphical representation of thermodynamic systems similar to circuit diagrams of electrical networks is adopted.

OUTLINE OF NETWORK THERMODYNAMICS ⸻⸻

In electrical network theory, one associates a current and voltage with each branch of the network. The current is required to satisfy a local conservation law, and the potential is required to be unique at each node of the network.

If *generalized effort* is denoted by $E^\star$, and the *generalized flow* is denoted by $F^\star$, the *generalized law of lumped energetic network* can be denoted as

$$E^\star = ZF^\star \tag{4.36}$$

where Z denotes the *generalized impedance*. The product

$$E^\star F^\star = P^\star \tag{4.37}$$

is called the *generalized power*, $P^\star$, which is the rate of change of energy.

The generalized versions of Kirchhoff's laws may be stated as follows.

1. The algebraic sum of effort differences (voltage drops) across the three impedance elements (resistance, capacitors, and the inductors) and across effort or flow sources (batteries) is zero around any loop.
2. The algebraic sum of flows (currents) into a node from all of its branches is zero.

The loop law is valid for all closed loops and not just for the minimum closed loop of any region. This is illustrated by the statement that if one walks around a mountainous path, and returns to the same spot, the net change of elevation is zero, irrespective of other elevated spots one may have traversed during the trip. The significance of this with the exact differential nature of thermodynamic functions is obvious. A network of vascular blood vessels may be considered in the same manner as an electrical network, since mathematically they have flows and resistances. The same algebraic forms of equations are obtained. However, when energy storage elements are introduced into the networks, the behavior becomes dynamic and a transient settling time is necessary to reestablish equilibrium, after any disturbance (any change in forcing conditions). Flow is always opposed by resistance in any physical situation. The resistance is overcome by suitable effort, and the result is a continuous dissipation of energy, resulting in the production of heat.

Thus, the generalized law of Equation 4.36 becomes

$$E^\star = RF^\star \tag{4.38}$$

where R is called the resistance. When resistance is independent of effort, one has linear laws. The nonlinear situation is denoted as $E^\star = E^\star(F^\star)$. In the linear dissipative situation, the power dissipation is expressed as

$$P^\star = RF^{\star^2} = \{E^{\star^2}/R\} \tag{4.39}$$

An element stores (potential) energy when its effort level increases as a result of the influx of the flow variable. The voltage across a capacitor increases in proportion to the amount of electricity in coulombs that has flowed in, while the force in a spring increases directly with its displacement.

Thus, one defines a *generalized displacement*, $H^\star$, as the integral of the flow variable, $F^\star$, over time.

$$H^\star = \int F^\star \, dt \tag{4.40}$$

The law of potential energy element is

$$H^\star = CE^\star \tag{4.41}$$

where C is the *generalized capacitance,* that is, displacement per unit effort. Thus, the stored potential energy in an ideal element is

$$PE \text{ (potential energy)} = \int E^\star \, dH^\star$$
$$= \{H^{\star 2}/2C\} = \{CE^{\star 2}/2\} \tag{4.42}$$

The *generalized momentum, M,* is defined as the integral of the effort variable, $E^\star$,

$$M = \int E^\star \, dt$$
$$= LF^\star \tag{4.43}$$

where L is the *generalized inductance* (with units of effort per unit rate of change of flow). The kinetic energy stored in a *generalized linear element* can be expressed as

$$KE \text{ (kinetic energy)} = \int M \, dF^\star$$
$$= \{LF^{\star 2}/2\} = \{M^2/2L\} \tag{4.44}$$

Equations 4.42 and 4.44 are simply generalizations of more evident relations for potential and kinetic energies.

Kirchhoff's laws are restatements of conservation and continuity restrictions. Thus, these are independent of the nature of the elements comprising a network. They constitute a set of linear constraints on the instantaneous values that the variable can attain. The form of these constraints depends on the manner in which the network elements are connected. With respect to the subject of regulation and control in biology, although the dynamic modeling of biophysical systems using network thermodynamics is impressive, no concrete example where the method has yielded useful and new information can be cited.

IRREVERSIBLE THERMODYNAMICS AND CHEMICAL REACTIONS

The relevance of stationary-state coupling between chemical reactions and diffusion flows for biological processes has been recognized. Chemical reactions by their very nature bring about spatial inhomogeneities, whereas diffusion processes lead to spatial homogeneity. Therefore, stationary-state coupling between matter fluxes and chemical reactions contain two processes with conflicting tendencies. One should recall that the validity of Curie's theorem forbids the existence of such coupling between chemical reactions and matter fluxes in isotropic systems. Proponents of the phenomenological expression of such coupling counter that such coupling is permitted in biological systems since most membrane systems where cross effects due to chemical reactions and matter fluxes occur are inhomogeneous and anisotropic.

One of the basic objectives of a molecular theory is to find expressions for such phenomenological coefficients which are macroscopically observable in terms of the properties of the molecules. It should be recognized that the coupling between chemical reactions and matter fluxes arises by the natural demand of the constraints imposed by equations of conservation and continuity. The description of the influence of chemical reactions on fluxes of diffusing species can possibly be described by phenomenological expressions whether they are valid or not. But such a description is inadequate as far as the explanation of the processes, such as active transport in biological systems. The equation of continuity essentially expresses the coupling between the two irreversible processes.

The rate of entropy production due to a chemical reaction is written as

$$\sigma = J_r(\mathscr{A}_r/RT) \tag{4.45}$$

where the flow term is represented by the rate or velocity of chemical reaction J_r of the Rth chemical reaction, and the subscript r is the affinity of such a chemical reaction. Although contribution to the rate of entropy production due to chemical reaction is written in the spirit of the linear phenomenological theory in terms of conjugate fluxes and forces, it is usually a poor approximation to write reaction fluxes, velocities, as linear, homogeneous functions of the forces, ΔF, decrease in free energy.

The reaction velocity is not proportional to change in the Gibbs free energy of the chemical reaction, except when the reacting system is very close to equilibrium. In general the dependence of $J_r(\mathscr{A}_r)$ on affinity is of the form

$$\{1 - \exp(-\mathscr{A}_r/RT)\}$$

Equation 4.45 results when one expands the exponential in powers of $(\mathscr{A}_r/RT)$ and retains only the leading two terms. Neglect of third and higher-order terms are justified if and only if

$$(\mathscr{A}_r/RT) \ll 1$$

For most chemical reactions that we know of this ratio $(\mathscr{A}_r/RT)$ is greater than unity, or is of the order of unity. Thus, phenomenological coefficients defined on the basis of Equation 4.45 are a poor representation.

In striking contrast, other irreversible phenomena like diffusion are well described by linear laws over a relatively wide range of flows and forces. Therefore, even in an anisotropic system where one may presume that Curie's theorem is not valid, the expression of cross–phenomenological coefficients for coupling between chemical reactions and matter fluxes is at best questionable. The main difficulties in the formulation of a system of thermodynamics for biological processes are concerned with the difficulties of thermodynamics itself.

When a chemical reaction occurs in a system, it can be described as

$$\sum_{i=1}^{k} v_i \mu_i = 0 \tag{4.46}$$

where μ_i is the chemical potential of the reacting species i, and v_i is its stoichiometric coefficients, namely, the number of grams of component i produced or consumed per gram of the reaction. Since the mass of a reacting system is conserved, while the number of moles in such a system is not conserved, it is more convenient to use the gram basis rather than the mole basis, while discussing chemical reactions. The coefficient v_i is negative, if i is a reactant in the chemical reaction, while it is positive if i represents a product of the chemical reaction. If a component j is present in the system and does not participate in the chemical reaction, then corresponding v_j is zero.

As a result of the chemical reaction, the mass of the component i changes with time. If dm_i is the change in the mass of component i in a time interval, dt, due to the chemical reaction since the total mass is conserved one has

$$\sum_{i=1}^{n} dm_i = 0 \tag{4.47}$$

A progress variable ξ can be defined as the grams of reaction occurred per gram of the original reactants. With this definition

$$\{dm_i / v_i\} = m \, d\xi$$

$$m = \sum_{i=1}^{n} m_i \tag{4.48}$$

where m denotes the total mass of the system. Division of Equation 4.48 by the total volume of the system yields the expression in terms of mass densities. The change in progress variable d is independent of component i.

A procedure for the treatment of chemical reactions which tallies with the analysis of other irreversible processes was developed by Kirkwood and Crawford (1952). They decomposed the progress variable $d\xi$ into two terms:

$$d\xi = d\xi(o) + \eta \tag{4.49}$$

The parameter $d\xi(o)$ is the equilibrium value of the progress variable for the instantaneous local temperature, pressure, and composition. The variable η (for each reaction) is a measure of the deviation of the chemical reaction from its equilibrium. The equilibrium value $d\xi(o)$ is determined by the condition that the Gibbs free energy change for the reaction vanishes. Chemical lag can be considered in terms of its relaxation behavior.

The affinity of a chemical reaction is defined as

$$\mathscr{A} = \sum_{i=1}^{j} n_i \mu_i - \sum_{h=1}^{k} n_h \mu_h \tag{4.50}$$

where i denotes reactant species and h denotes the product species. Utilizing the simple ideal solution expression for the chemical potential of solute species,

$$\mu_i = \mu_i^o + RT \ln C_i \tag{4.51}$$

Equation 4.51 can be written as

$$\Delta G = -\Delta G(o) + RT \sum_{i=1}^{j} n_i \ln C_i$$
$$- RT \sum_{h=1}^{k} n_h \ln C_h \tag{4.52}$$

$\Delta G(o)$ is the weighted difference in standard free enthalpies, corresponding to the equilibrium constant K of the reaction,

$$G(o) = -RT \ln K \tag{4.53}$$

For a reaction of the kind

$$A \underset{k_1^*}{\overset{k_1}{\rightleftharpoons}} b$$

the expression for affinity and chemical flow can be written as

$$\begin{aligned} \text{affinity:} &\quad RT \{\ln k_1 a - \ln k_1^* b\} \\ \text{flux:} &\quad k_1 a - k_1^* b \end{aligned} \tag{4.54}$$

We have utilized the convention that small letters denote concentrations of species concerned. (It should be noted that all these expressions suffer from the approximation that reaction velocity is proportional to affinity. The treatment of chemical reactions either alone or in conjunction with other irreversible processes has not yet reached a satisfactory state in irreversible thermodynamics.

Similarly, for an association–dissociation reaction of the kind

$$A + B \underset{k_1^*}{\overset{k_1}{\rightleftharpoons}} C$$

one has

$$\text{affinity:} \quad RT\{\ln k_1ab - \ln k_1^\star c\}$$
$$\text{flux:} \quad k_1ab - k_1^\star c \tag{4.55}$$

Finally, for a reaction of the kind

$$A + B \underset{k_1^\star}{\overset{k_1}{\rightleftharpoons}} C + D$$

one has the results

$$\text{affinity:} \quad RT\{\ln k_1ab - \ln k_1^\star cd\}$$
$$\text{flux:} \quad k_1ab - k_1^\star cd \tag{4.56}$$

The chemical flux defined in this manner has the dimension of concentrations (moles/unit volume) per unit time. Since the thermodynamic force responsible for driving the chemical reaction has the dimension of energy per mole, it follows that the dissipation function should have the dimension of energy per unit volume per unit time. The expression for the chemical potential of species as a function of its concentration is valid only very approximately in biological systems.

Similarly, for matter fluxes which have the dimension of moles per unit area per unit time and for the corresponding thermodynamic force, namely, the gradient of chemical potential with the dimensions of energy per mole per unit length, the dissipation function will again have the dimension of energy per unit volume per unit time. The practice of expressing differences in chemical potential as the thermodynamic force for matter fluxes leads to dimensional inconsistencies.

In the language of chemical kinetics, self-replication is considered an autocatalytic process. Both in catalytic and autocatalytic reactions, the same species appear as a product. In catalytic reactions, the flux is proportional to the concentration of the catalyst, which does not change with time. In the case of autocatalytic reactions, the flux is proportional to net production of the catalyst. The rate of change of concentration with time is zero, when the concentration of the catalyst is zero. It is also zero when the reaction is at equilibrium. Therefore, it follows that there be an extreme, maximum, of concentration between these two extreme situations. The simplest example of autocatalysis is represented by the second-order chemical reaction:

$$A + B^k \rightarrow 2A$$
$$\{da/dt\} = \dot{a} = kab \tag{4.57}$$

Under conditions when the concentration of B is maintained constant (for example, $b = b(o)$ from an external supply), the rate at which concentration of A increases is proportional to concentration of A. The solution profile of the differential equation is evidently an exponential function,

$$a(t) = a(o) \exp\{kb(o)t\} \tag{4.58a}$$

where $a(o)$ is the initial concentration of a at time $t = 0$. This kind of exponential growth resulting from autocatalysis is possible only far off from equilibrium. Close to equilibrium, one observes always a relaxation behavior. It should be noted that the reaction presented in Equation 4.57 is unidirectional. Including the reverse reaction,

$$A + B \underset{k_1^*}{\overset{k_1}{\rightleftharpoons}} 2A$$

$$\{da/dt\} = \dot{a} = k_1 ab - k_1^* a^2 \tag{4.58b}$$

Therefore, increase of concentration of A with time exhibits a completely different behavior when the reaction is closer to equilibrium and when it is far off from equilibrium. These are exhibited in Figure 4.1.

The equations for the generalized forces and fluxes from an irreversible thermodynamic point of view can be written as

$$\text{flux:} \quad J = \dot{a} = k_1 b(o)a - k_1^* a^2$$
$$= a\{k_1 b(o) - k_1^* a\} \tag{4.59}$$
$$\text{force:} \quad X = \{\mathscr{A}/T\} = -R \ln\{a/a_{eq}\}$$

Close to equilibrium, one has the proportionality between flux and force, as required by the linear phenomenological relations of irreversible thermodynamics, namely,

$$\text{flux:} \quad J = \dot{a} \sim k_1^* a_{eq}[a_{eq} - a]$$
$$\text{force:} \quad X \sim \{R/a_{eq}\}[a_{eq} - a] \tag{4.60}$$

where a_{eq} denotes the concentration of A at equilibrium of the chemical reaction. Evidently, these relations are only approximate.

Equation 4.60 demonstrates an interesting and unique aspect of autocatalytic reaction. The entire range of physically accessible concentrations may be subdivided into two ranges, as presented in Figure 4.2. These are the range I around equilibrium and at concentrations of A larger than its equilibrium value a_{eq}, and the range II covering small concentrations of A up to a certain critical value. In range I, both functions $J(a)$ and $X(a)$ have negative first derivatives. Chemical fluxes and affinity decrease as one approaches equilibrium.

A different behavior is observed in range II. The flux increases as the force decreases. An increase in the concentration of the reaction product causes an increase in the reaction rate, which in turn leads to more product. Thus, an instability arises. Autocatalysis is a necessary and sufficient condition for the occurrence of such a range of instability. Other

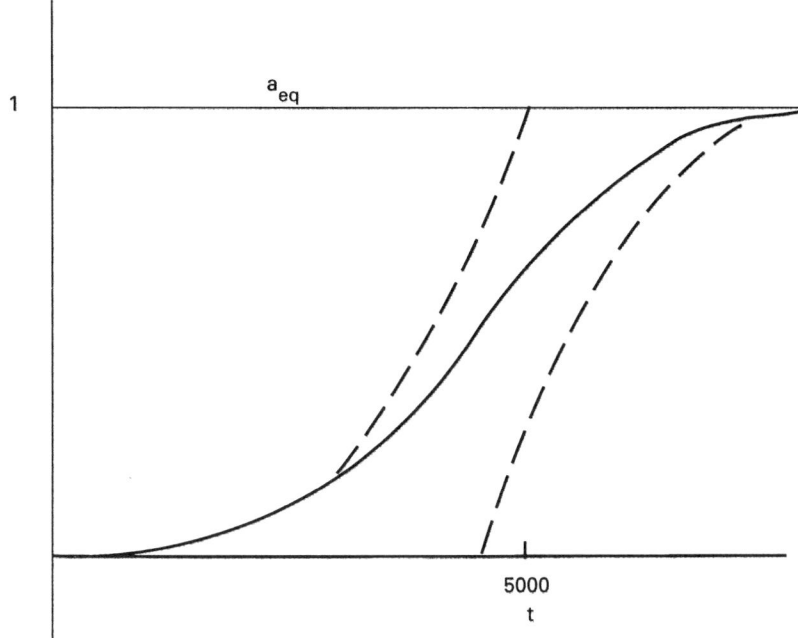

Figure 4.1 A characteristic solution curve of concentration of A, at buffered concentration of B, $b = b_0 = b_{eq}$.

$$a(t) = a(o) \exp[kb_0 t].$$

Pure exponential growth and relaxation to equilibrium are shown in dotted lines. $f_1 = 1$, and $f_2 = 0.001$ in $\{tc\}^{-1}$, b_{eq} in $0.001\,c$ in arbitrary units of time and concentration. (Reproduced with permission from Springer-Verlag, "Biophysics," chapter 17.2, Peter Schuster.)

elementary chemical reactions in which the flux is proportional to one of the reaction products do not meet the conditions of instability, resulting in increase of flux as force decreases. (It should be noted that an explanation for this negative force-flux relation existence does not come from this argument. Only that such relations are permitted to occur in steady state is emphasized by these arguments.)

Schuster (1982) has presented the conditions under which a chemical reaction can be kept far off from equilibrium. He concludes that it is not sufficient to introduce a recycling process in order to keep the autocatalytic reaction far off from equilibrium. One needs an irreversibility step to keep the reaction within the range of a negative force-flux characteristic of range II. In range II, the flux increases as the force decreases (see Figure 4.2). This is an analogy to the negative current-voltage characteristics observed in electronic tubes and reflects an inherent instability. However, no explanation

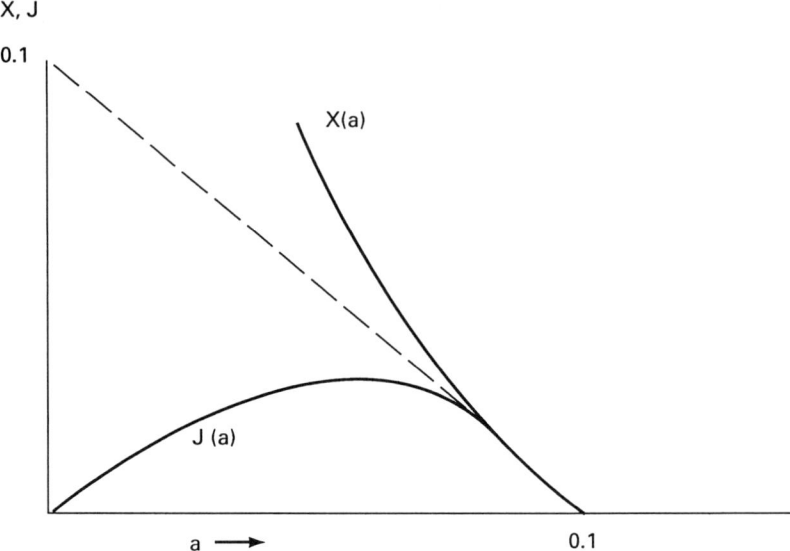

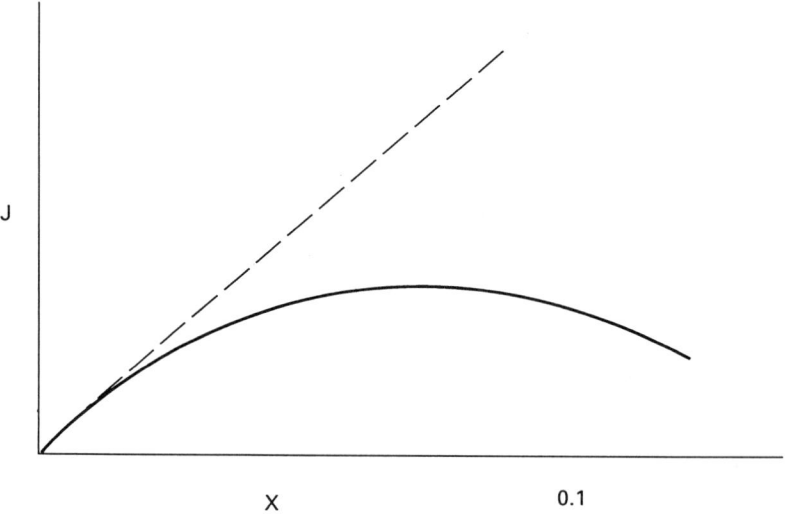

Figure 4.2 Relations between chemical forces $X(a)$ and Chemical fluxes $J(a)$ in an autocatalytic reaction, $A + B \; 2A$, denoted in Equation 4.39, at buffered concentration of B. Units are chosen such that the tangent at equilibrium part, $a = a_{eq} = -1$. $k_1 = 1$ and $k_1^\star = 0.001$. (Reproduced with permission from Springer-Verlag, "Biophysics," chapter 17.2, of Schuster.)

is available as to why a negative force-flux relation should arise or even exist. In general, biochemical reactions follow complicated multistep mechanisms, which are difficult to analyze. The corresponding simplified overall rate equations do not follow mass–action kinetics.

It is safe to conclude that the influence of chemical reactions on fluxes of species is latent in the equations of continuity and mass conservation relations. The magnitude of flux and its direction of a specified species undergoing transport is determined by the magnitude of force acting on it and the resistance offered to transport of this species by the medium. The equation of continuity stipulates that fluxes of different species are coupled, and that one species cannot move independently of another. This is reinforced by the existence of intermolecular interactions which act over distance. Thus, the magnitude and direction of flux of a specified species is determined in an involved manner, by all the forces acting on species of the system, as shown briefly by the following set of equations.

Under stationary-state conditions that are isothermal and isobaric in nature, the Gibbs–Duhem relation (for one dimension)

$$\sum_k C_k(x)\{d\mu_k/dx\} = 0 \tag{4.61}$$

where the summation should be performed over all species present in the system. This is the force-balance equation valid for steady state. The expression for the gradient of chemical potential, that is, the mean force acting on species k, as given by molecular theory, (Vaidhyanathan, 1977, 1980) is

$$\{d u_k/dx\} = -[J_k/C_k(x)]\left\{\sum_\xi C_k\xi_{kj}\right\} + \sum_j J_j\zeta_{kj} \tag{4.62}$$

The frictional coefficient ξ_k that a molecule of kind k experiences in the system (diffusion barrier) at a specified location x can be written as

$$\xi_k = r_k + \sum_j C_j(x)\xi_{kj} \tag{4.63}$$

where r_k is the contribution to the subscript k arising from fixed molecules of the membrane framework, through intermolecular interactions. The subscripts kj are known as the partial frictional coefficients and represent the contribution of molecules of kind j to the frictional coefficient of molecule of kind k. The frictional coefficient is related to the diffusion coefficient of molecule k, D_k, by the relation due to Einstein,

$$D_k = [k_B T/\zeta_k] \tag{4.64}$$

where k_B is the Boltzmann constant, and T is temperature in the absolute scale.

A second expression for the gradients of chemical potential, obtainable from Equation 2.35 is

$$\{d\mu/dx\} = k_B T\{d \ln C_\sigma(x)/dx\} + Z_\sigma e\{d\phi/dx\}$$

$$+ He^2 \sum_\sigma \sum_\eta Z_\sigma Z_\eta \{dC_\eta/dx\} \qquad (4.65)$$

$$+ H^\star Z_\sigma e \sum_j \{dC_j/dx\}$$

where j denotes nonionic species and σ and η denote ionic species. e is protonic charge. The terms H and $H^\star$ of Equation 4.65 are the molecular integrals, contributing to chemical potentials of ionic and nonionic species in solutions. These are defined and identified in Appendix 3B.

Fick's law demands that the flux of a species be in a direction along the decrease in concentration of the species. The stability considerations require that the diffusion coefficients, which are the diagonal elements of the diffusion matrix, be positive definite. Thus, if one believes in the simple diffusion expression of Fick's law, then observed phenomenon like active transport cannot be explained in this context. However, recognition of the existence of coupling between fluxes and forces, within or beyond the validity of linear phenomenological relations, permits one to find explanations for active transport. The partial frictional coefficients, which are the off-diagonal elements, can be positive or negative, depending on the nature of intermolecular interactions. Thus, the effective resistance suffered by a molecule during transport in a membrane system, as denoted by the frictional coefficient, can change its magnitude, as well as its sign, depending on the system and its component interactions. One suspects that herein lies the basic mechanism for active transport and ion pumps.

We now turn our attention to an allied subject of catastrophy. It is well known that in the deterministic description when all fluctuations are neglected, a nonlinear system may be in a variety of stable stationary states. The transitions between different branches of stationary states arising due to instabilities of the system is of interest. Catastrophy theory can be very useful in classifying the different instabilities occurring in dynamic systems, under the influence of fluctuations.

CATASTROPHY

Cells which may have been exhibiting normal growth become cancerous and multiply rapidly. A catastrophe is usually associated with unpleasant consequences. The final word in everyone's biography is death. Thus, catastrophy can denote the sudden abrupt change of state which may or may not have desirable features. Such discontinuous changes of state are analyzed mathematically on the basis of catastrophy theory. In this section, an outline of the introductory aspects of catastrophy theory is presented.

Externally driven nonlinear systems exhibit various behaviors, depending on the value of internal or external parameters. The system undergoes continuous or sometimes abrupt changes from one state to another, for some critical value of the parameters. The determining parameter is called the bifurcation parameter. There are mainly three kinds of bifurcations (Kaiser, 1988): (1) Hopf bifurcation, (2) saddle–node or tangent bifurcation, and (3) period–doubling bifurcations. The period–doubling bifurcation, which has received more attention, represents the case of an oscillation (limit cycle of period T) bifurcating into another limit cycle of oscillation with period $2T$. This can further result in additional bifurcations of periods $4T$, $8T$, and so on. Additional remarks about bifurcations are presented in Chapter 6. For our current purpose, it is sufficient to remark that bifurcations can lead to chaotic motions.

A state of a system can be expressed in terms of the variables, say, $x_1, x_2, \ldots, x_n$. These variables could be concentrations of chemicals in the cell, for example. Changes in the value of these variables with time t permit description of the evolution of the system. One may require that small variations in the values of these state variables lead only to small alterations in the configuration of the system and that large-scale effects due to small changes are not permitted.

In addition to these state variables, there could be a set of interaction parameters y_i's which characterizes the strength of interaction between various parts of the system. A change in the environment can lead to deviations in the values of y_i's from their original values, such changes being small. Two possibilities now exist. Either the state of the system will be changed a little, or it will change markedly. In the second case, a catastrophy occurs. The set of values $y_1, y_2, \ldots, y_k$ for which this marked change occurs is called the *catastrophy set*. The parameters y_i's are called the *system parameters*, and the x_i's are called the *system variables*. The state function, $V(x_i\text{'s}, y_i\text{'s})$, such as energy or the Hamiltonian or some other potential function, is determined by the state variables and the system parameters.

When some extremization condition is satisfied, such as the minimization of potential energy, the system evidently is in an equilibrium state. This n extrema condition can be denoted as

$$\{\partial V/\partial x_1\} = 0; \quad \{\partial V/\partial x_2\} = 0; \cdots \quad \{\partial V/\partial x_n\} = 0 \quad (4.66)$$

The solution of Equation 4.66 depends on the system parameters. When these vary smoothly with $y_1, y_2, \ldots, y_k$, there is regularity. If these do not vary smoothly, then there is catastrophy.

If due to a small disturbance, y_i changes its value from y_i to $y_i + K_i$, for $i = 1, 2, \ldots, k$ and x_j becomes $x_j + X_j$ with $j = 1, 2, \ldots, n$, in the new time-independent state, one can expand the new time-independent state in terms of the old time-independent equilibrium state in Taylor expansion.

$$\{\partial/\partial x_j\}V[x_1 + X_1, \ldots, x_n + X_n, y_1 + K_1, \ldots, y_j + K_j]$$

$$= (\partial/\partial x_j)V[x_1, \ldots, x_n, y_1, \ldots, y_j]$$

$$+ X_1\{\partial^2/\partial x_1\,\partial x_j\}V[x_1, \ldots, x_n, y_1, \ldots, y_j\} + \cdots$$

$$+ X_n\{\partial^2/\partial x_n\,\partial x_j\}V[x_1, \ldots, x_n, y_1, \ldots, y_j]$$

$$+ K_1\{\partial^2/\partial y_1\,\partial y_j\}V[x_1, \ldots, x_n, y_1, \ldots, y_j] + \cdots$$

$$+ K_j\{\partial^2/\partial y_k\,\partial x_j\}V[x_1, \ldots, x_n, y_1, \ldots, y_j] + \cdots$$

$$(4.67)$$

The first term vanishes identically because of the equilibrium conditions in Equation 4.66. Thus, for the new time-independent state, the variations $X_1, X_2, \ldots, X_n$ must satisfy the conditions when higher-order terms are neglected,

$$X_1\{\partial^2 V/\partial x_1\,\partial x_j\} + \cdots + X_n\{\partial^2 V/\partial x_n\,\partial x_j\}$$

$$= -K_1\{\partial^2 V/\partial y_1\,\partial x_j\} - \cdots - K_k\{\partial^2 V/\partial y_k\,\partial x_j\} \qquad (4.68)$$

$$\text{for all} \quad j = 1, \ldots, n$$

Equation 4.68 expresses the set of n linear equations for the n unknowns $X_1, \ldots, X_n$. If the determinants of the coefficients on the left-hand side is nonzero, there exists a unique solution, which gives each X_i in the form of a linear relation involving K_i's and some constant parameter coefficients. The smaller the K_i's are, the smaller is the variation X_i and the time-independent state resulting due to environmental perturbation is regular.

On the other hand, when the determinant is zero, a general solution of Equation 4.68 contains an arbitrary element, which does not depend on $K_1, \ldots, K_j$, since this satisfies Equation 4.68 with zero on the right-hand side. Therefore, X_i cannot be forced to become zero, with $K_1, \ldots, K_j$. One has in this situation, a catastrophy.

Therefore, one has the theorem: If $y_1, \ldots, y_k$ are such that an equilibrium state also satisfies the condition that the matrix of the second-order variations of the potential functions with state variables is singular, then the state parameters $y_1, \ldots, y_k$ form a catastrophy set. If the determinant of such a matrix (Hessian of V) does not vanish, then the time-independent state is regular. It is implied that external circumstances affect the system parameters. The theorem can be restated that an equilibrium state is a catastrophic or a regular state according to whether the Hessian of the potential function does or does not vanish. If a system is an equilibrium state and is regular, and the state variables are altered fractionally while the state parameters are fixed, will the state stay near equilibrium or tend to move away from it? In this situation the criterion is that if for every small change in the x_i's, the potential V increases from its equilibrium value, the equilibrium state is said to be stable. If for small variations in the x_i's there is no increase in V, then the equilibrium state is called unstable.

In this chapter, we have outlined some of the problems needing to be solved by thermodynamics of biological processes. There exists among some scientists the belief that organisms function normally, and that their functioning can be described by linear thermodynamic relationships. It is also assumed that in general, processes occurring in living systems are regulated by control systems that prevent the organism from entering the nonlinear region. The probability of the occurrence of giant fluctuations moving the system into the nonlinear region is only an exceptional case.

REFERENCES

BLUMENTHAL, R., and A. KATCHALSKY, *Biochim. Biophys. Acta,* vol. 173, (1969), p. 357.

FITTS, D.D., *Nonequilibrium Thermodynamics.* New York: McGraw-Hill, 1962.

JONES, D.S., and B.D. SLEEMAN, *Differential Equations and Mathematical Biology.* London: George Allen and Unwin, 1983.

KAISER, F., in *Biological Coherence and Response to External Stimuli,* p. 25, ed. H. Frohlich. Berlin: Springer-Verlag, 1988.

KIRKWOOD and CRAWFORD, *J. Phys. Chem.,* vol. 56, (1952), p. 1048.

KIRKWOOD, J.G., and I. OPPENHEIM, *Chemical Thermodynamics.* New York: McGraw-Hill Book Co., 1961.

KREUZER, H.J., *Nonequilibrium Thermodynamics and Its Statistical Foundations.* New York: Oxford University Press, 1981.

LAMPRECHT, I., and A.I. ZOTIN, *Thermodynamics and Kinetics of Biological Processes.* New York: de Gruyter, Publishers, 1983.

MEIXNER, J., *J. Math. Phys.,* vol. 4, (1963), p. 154.

SCHUSTER, P. *Biophysics.* Springer-Verlag, 1982.

TRUESDELL, C., *Rational Thermodynamics.* New York: McGraw-Hill Book Co., 1969.

VAIDHYANATHAN, V.S., and M.S. SESHADRI, *Biochim. Biophys. Acta,* vol. 211, (1970), p. 1.

VAIDHYANATHAN, V.S., *J. Theor. Biology,* vol. 31, (1971), p. 53.

VAIDHYANATHAN, V.S., in *Topics in Bioelectrochemistry and Bioenergetics,* vol. 1, p. 288, ed. G. Millazzo. New York: John Wiley and Sons, 1977.

VAIDHYANATHAN, V.S., in *Ions, Surfaces, Membranes Advances in Chemistry Series,* vol. 188, p. 20, ed. M. Blank. American Chemical Society, 1980.

YOURGRAU, W., A. VAN DER MERWE, and G. RAW, *Treatise on Irreversible and Statistical Thermophysics.* New York: The Macmillan Company, 1966.

chapter 5

Biochemical Control Mechanisms

INTRODUCTORY REMARKS

The thrust of modern biology is toward investigations of increasingly complex chemical reactions occurring in compartments of heterogeneous subsystems. In biology, control at the cellular and molecular levels is evidently accomplished by the regulation of concentrations of species involved in chemical reactions that are occurring or are needed to occur. In order that cells of an organism act in the coordinated fashion necessary for efficient functioning, they must be able to receive and transmit information. Such information transfer is believed to be mediated by molecules released by the cells. The action sites of such molecules could be local, as in the case of neurotransmitters, or long range, as in the case of hormones. Regardless of the range of interactions, a cell must be able to distinguish the signals relevant to it from an enormous flow of noise with a high degree of precision.

Control requires a source of information which may be either intrinsic or extrinsic. Intrinsic control is built into the system as the regulatory effects of mass-action law and Le Chatelier's principle, in the primary metabolic control of the energy-processing component. Extrinsic control is imposed on this ergonic system by the nucleic acids, or by the neurosensory effector, or the endocrine factors. It is known that there are multiple patterns of metabolic control, interactions of which have been described as concerted, cooperative, and cumulative feedback inhibition.

At the molecular level, the study of control and regulation involving the conjugate concepts of structure and function is the central theme of biophysics. Any sudden, continuous, or periodic changes of the physico-chemical parameters of a metabolic system result in perturbations of the stationary state which is followed by reestablishment of another time-independent stationary state. The time required for the system to attain this new stationary state is called the *characteristic relaxation time*. The relaxation time is specific for the system and is related to the response time and transfer function of the system. Due to the complexity of the metabolic systems in which the subsystems with different control characteristics interact, control characteristics of slow processes become time independent, and these dictate the time structure of the total system. The complexity of a transient is based on the nonlinearity of the metabolic reactions. The key to our understanding of the complex and cell–specific motion of the metabolic processes lies in the investigation of the time evolution of such reactions.

The conditions inside the cell depict the prevalent microenvironments, and these reflect only partially the whole-body conditions. The observation that concentrations of certain chemical species are maintained in cells at relatively constant values is at best only circumstantial evidence for the existence of a regulatory system. The importance of a postulated control mechanism to physiology should be demonstrated prior to it being accepted as an essential reaction.

Any chemical kinetic model should be consistent with the conservation relations, nonnegativity of concentrations and temperature, and satisfy the requirements of the laws of thermodynamics. The first law of thermodynamics describes the energy equation and balance. The requirement that the kinetic model satisfies the second law of thermodynamics is somewhat involved, since this requires prior knowledge of the entropy and chemical potentials, which are not available for all systems.

Many chemical reactions occur within the cell, and most of them are catalyzed by enzymes. The network of biochemical reactions with a single cell is controlled and coordinated by an array of control and feedback pathways, the diversity of which is far greater than one may comprehend. Regulation of these chemical reactions is achieved by the modulation of key reactions that control metabolic fluxes. This is accomplished by either controlling the quantity of the enzyme or by the control of the catalytic activity of a given amount of the enzyme. Some of these chemical reactions can be described by linear differential equations. But most of these biochemical reactions are nonlinear, resulting in a variety of response patterns. In such situations, the behavior of the reactions is similar to analysis of multicompartment model systems.

The mechanisms by which biological processes take place in nature should be explained ultimately in the transformation of one chemical species into another. Therefore, chemical kinetics related to the synthesis and consumption of chemical species occurring in open systems is important. Chemical kinetics is concerned with the determination of the rates and mechanisms of reactions. The concept of the reaction mechanism relies on the concept of an elementary reaction. This is a reaction that corresponds to a single molecular collision. A reaction is generally composed of a number of elementary step reactions which as a whole constitute a reaction mechanism. Open systems do not in general possess thermodynamic potential functions, a fact which manifests itself with the existence of more than one time-independent (stationary) state. Open systems still satisfy the conditions of positive entropy production. Empirical rate functions will not in general be consistent with thermodynamics. The asymptotic behavior in this situation can be analyzed by means of the Liapounov functions having the properties of a postulate. These nonthermodynamic Liapounov functions would also be useful when thermodynamic expressions for the entropy are not available.

Molecular recognition at the cell surface is mediated by receptors. Typically, receptors bind to the ligands they are designed to recognize with affinities of the order of 10^8 liter per mole, and they translate that interaction into a sequence of signals that ultimately leads to biological activity. The transfer of information that leads from binding to any particular type of activation often involves a large number of biochemical steps. It is not always known whether the information that is transferred resides in the receptor, in the ligand, or in the complex. One possible mechanism for signal transduction at the cell surface is ligand-induced receptor clustering. Such clustering would be facilitated by a multivalent ligand. Insulin-dependent glucose oxidation and the activation of the epidermal growth factor appear to require at least local aggregation of appropriate receptors (DeLisi, 1980). A conformational change in the receptor would be a mechanism for negative cooperativity. The binding of a ligand to a receptor site is negatively cooperative, if its binding free energy is less favorable when an adjacent site is occupied than when it is unoccupied.

Regulatory molecules present in small amounts bind numerous enzymes, changing their shape to one required for catalysis on a particular substrate. The regulators (allosteric effectors) are not utilized in the reaction and thus, even though they are present in very small amounts, they can exert enormous influence on regulation. Another possible reason for accelerated dissociation that would be biologically significant is ligand-induced clustering. When ligands are clustered in space, competition among the receptors for ligands may occur. Such competition can lead to different kinetics of ligand receptor binding and dissociation can be quite different for receptors in clusters compared to receptors uniformly distributed in solution.

Biochemical processes appear to use all possible kinds of enzyme control, namely, control of enzyme effectiveness by activation and inhibition, and control of enzyme concentration. The processes of induction and repression serve to enhance or retard the rate at which the enzyme is synthesized within the cell and thus indirectly control concentrations. Since enzyme molecules are continually being destroyed or lost to the environment by the cell by a number of processes, these enzyme molecules must be synthesized continually at a rate necessary to maintain the normal complement. The importance of enzymes in biological systems is that they make it possible to establish equilibria very fast, while in their absence most metabolic reactions would proceed very slowly. The modulation of the catalytic activity of enzymes occurs through several kinds of mechanisms. Conformational changes which alter the interactions between polypeptide chains as well as through changes in aggregation states are of importance for the regulation of enzymes.

Control of the rate at which enzyme molecules are synthesized becomes a usable control mechanism only because there is a normal turnover. Repression is thus a reduction in the rate of synthesis, so that normal losses of the enzyme will reduce the concentration of those molecules remaining in the cell. Induction is presumed to be a derepression or interference with the repression mechanism. Whereas activation and inhibition provide a fine control of the enzyme-catalyzed chemical reactions, induction and repression provide a relatively crude or coarse control.

Enzymatic activities are often switched on or switched off, increased or decreased by the physiological environment. The binding of small molecules such as substrates or effectors to enzymes is an important mode of regulation. Regulation of such activity can be accomplished in possibly three ways. (1) Competitive regulation and allosteric regulation affect the enzyme molecules by either chemical or structural means. (2) Since many enzymes have distinct catalytic and regulatory subunits, effector molecules can directly act on the regulatory subunits, which in turn influence the activity of catalytic subunits. (3) At the physiological level, control may be effected also by changes in the rate of biosynthesis of the enzyme.

A number of chemical reactions occur in a biochemical system sequentially. The identification of one out of a number of processes operating in this manner, which in effect sets the pace for the entire system of reactions, is useful when considering the control aspects of a biosynthetic pathway. The concept of rate-controlling or rate limiting has a somewhat different meaning in the stationary states. If one of the reactions is controlled by an enzyme, the rate of conversion of one species into another may or may not obey the law of mass action. When mass-action law is not valid, then this catalyzed step is rate limiting in the steady state. An enzyme-controlled reaction may thus serve as a controlling reaction for an entire sequence of processes, if the operation is in the region of substrate saturation.

For any particular sequence of steps in an enzymatic reaction, it is possible to write down a set of simultaneous equations, one stating that the sum of the concentrations of all forms of the enzyme remains constant, and one for each enzyme complex, stating that its concentration remains constant in the steady state, expressed in terms of the rate constants involving that complex. It is possible by standard methods to solve this set of equations to give expressions for the concentration of the chosen intermediate.

The reaction rates are constant in steady states. This is called the kinetic behavior of the process. Its characteristics depend on the various constraints placed on the supply of the enzyme and its substrate. The dynamic behavior will occur when the system adjusts itself to a change in these constraints such as might be occasioned by a change in the total amount or effectiveness of the enzyme brought about by some control action.

For a chemical reaction which has a series of intermediate steps, if the initial reactants and the final product concentrations are maintained at fixed values, one expects that for a given set of values of concentrations of reactants and products the concentrations of the intermediates will assume time-independent well-defined values in the steady states. The equilibrium situation can be obtained by the constraints imposed by the law of mass action. However, if in general one has a nonlinear scheme of reactions, there may be different time-independent solutions for the kinetic equations, all satisfying the conditions that concentrations be positive definite and time independent. One of these solutions may correspond to the chemical equilibrium case. The only requirement for the correctness of the solution is that it is stable with respect to perturbations. In order to prevent these systems from collapsing into an equilibrium state, they must be continually supplied with free energy, or with energy-rich matter. Living systems exhibit such properties.

In general, a metabolic control theory must be based on the cooperative properties of the enzymes as the control unit of metabolism as well as the feedback circuitry of the metabolic control net. The identification of the chemical processes and pathways comprising the regulatory circuits in biology is tedious. The individual processes must be identified, along with their input-output relations and the pathways from one process to another with all the inputs and outputs being located. Often, many equivalent mechanisms can account for a set of observed data in any system of chemical reactions.

A minimal organizational requirement for biological control is that the variable to be controlled is affected by two antagonistic processes, one which increases or activates the element and another which decreases or inhibits it. The antagonists may be forward and backward reactions of a reversible metabolic step or of different pathways affecting the same component in opposite directions. The antagonists at other levels could be excitatory and inhibitory neurons converging on a neuron pool in the central nervous system. These initial remarks evidently suggest that one should examine carefully the known results from chemical kinetics and kinetics of enzyme-catalyzed reactions.

CHEMICAL KINETICS

Activation is a very general process in nature and is a necessary prerequisite in a number of chemical reactions. The requirement of acquiring additional energy restricts the number of so-called *fruitful collisions* from the number of actual collisions occurring between molecules that chemically react in condensed systems. Catalysis of biological reactions is achieved by enzymes. Almost all biological reactions are associated with a specific enzyme. High catalytic power in conjunction with strict specificity con-stitutes the most striking feature of all enzymes. Since proteins are relatively stable molecules with a lifetime of the order of several minutes, while catalysis is a relatively fast process, the concentrations of proteins can become substantially large in the system. In order to control proliferation of proteins, the organism sometimes utilizes certain substances to repress the synthesis of such macromolecules.

The term *order* has a different meaning when dealing with chemical reactions as in kinetics than when it refers to the order of a differential equation. A second-order chemical reaction leads to a nonlinear differential equation, containing products of dependent variables, and whose solution bears no similarity to solutions of second-order differential equations. In chemical kinetics, the term *order* refers to the number of molecules that combine to form a new molecule. The rate of the reaction is proportional to to the products of concentrations of the two precursor molecules. This is because of the requirement that in order to chemically react, molecules must come close together, and the number of collisions that are essential for this to occur is obviously proportional to the number of reactant molecules (hence, concentrations) present in the system. This is the basis of the well-known law of mass action. The mathematical problem of systems and cycles of chemical reactions is related to the difficulties of obtaining an analytical form for the solutions of such a system of nonlinear differential equations. The mathematical analysis of certain rate processes, for which the roles of stimulus and initial conditions are analyzed, is all based on the equations for conservation of mass and the mass–action law for chemical reactions.

A stationary state is by definition one whose description does not change with time. This definition implies that it becomes impossible for a closed system undergoing chemical reaction to ever achieve a stationary state, since the composition of such a system is constantly changing with time. The concept of the *quasi-stationary state* is utilized in discussing sequences of consecutive reactions occurring in closed systems. Quasi-stationary state means a state whose instantaneous description differs from the truly stationary state by quantities which are negligibly small.

The occurrence of a detectable amount of intermediates in reactions catalyzed by enzymes is equivalent to the failure of the steady-state assumption frequently utilized in enzyme kinetics. The system cannot be represented any more by one single rate equation but should instead be described by a whole system of differential equations involving second-order rate terms. The steady-state analysis yields information about the overall rates rather than about single steps. In general, neither equilibrium nor kinetic studies can verify the validity of any mechanism. The only way of proving a mechanism is that if there are n possibilities, and if one can exclude $(n - 1)$ of these as nontenable, then the remaining one may be considered as proven. The detailed structure of chemical reactions within systems and subsystems can sometimes be described using graphical models or block diagrams of all interactions.

A metabolic system can be represented by the conversion of a substance X into another substance Y. In this case, the system represents a chemical transformation of X, usually catalyzed by an enzyme. The conversion of X into Y may represent several consecutive steps. Such a reaction may be influenced by temperature, pressure, and concentrations. For most metabolic systems, however, one may ignore the dependence on pressure and temperature, since these are essentially constants. The only significant factor influencing the chemical reaction is the concentrations of the intermediates. If the system in which the chemical reaction occurs cannot exchange X or Y with the outside, that is, no material is exchanged with the surroundings, it is called a *closed system*. For a closed system, equilibrium is the only state in which the concentrations of the intermediates do not change with time.

Most metabolic systems, however, do not represent closed systems. They permit both the supply of X and removal of Y from the system. An open system can exist in a stable state which is not an equilibrium state. Although a steady state provides a useful approximation to most metabolic systems, there are examples of nonsteady-state systems. For example, the concentrations of some glycolytic intermediates may oscillate under suitable conditions.

The law of mass action states that the reaction rates are proportional to the concentrations of the reacting species. In a closed system, there are no sources or sinks for the loss of material during the reaction. The situation of a reversible first-order chemical reaction occurring in a closed system is described by two first-order differential equations:

$$A \underset{k_1^{\star}}{\overset{k_1}{\rightleftharpoons}} B$$

$$\{da/dt\} = -k_1a + k_1^{\star}b$$
$$\{db/dt\} = k_1a - k_1^{\star}b$$

$$(5.1)$$

We utilize the convention that small letters denote the concentrations of species identified by capital letters. Thus, a and b represent respectively the concentrations of species A and B in the system. The system of differential equations in Equations 5.1 are the same as one would obtain for a two-compartment model-system, in which one presumes that A and B are transferred between the compartments. Also adopted is the convention that the rate constant of a reverse step of a reaction i with a forward rate constant k_i is denoted by $k_i^\star$. If in the system at time $t = 0$, a and b equal zero, and if the introduction of species A changes the value of a from zero to a^o, one has the initial conditions

$$a = 0 = b \quad \text{for} \quad t < t_0$$

$$a = a^0; \qquad b = 0, \quad \text{for} \quad t = t_0$$

The situation of a closed system in which a reaction represented by Equation 5.1 occurs is schematically presented in Figure 5.1. The system evolves toward a steady state when ultimately a will equal its stationary-state value $a^\star$ and b equals $b^\star$. Since the system is closed, the response curves are single exponentials so that this is a first-order dynamical system. The time constants are different for the two transients. The final steady-state values depend on the total material present, but the ratio $\{a^\star/b^\star\}$ is solely determined by the rate constants.

$$\{a^\star/b^\star\} = [k_1/k_1^\star] \tag{5.2}$$

Assume that a disturbance (perturbation) occurs at time $t = t_1$, when the rate constant k_1 is set arbitrarily equal to zero. Though such a change may not be realizable in practice, such a concept provides further insight into the process behavior. The concentrations return to their initial values with longer time constants.

The dynamic behavior of a given system with chemical reactions may be represented in either the time domain or the frequency domain. The response of the system to a given disturbance has been represented as a function of time, or somewhat indirectly in terms of the sinusoidal behavior and its frequency spectrum. Another mode of representation called the *phase-plane representation* represents the response to a specified disturbance and the initial conditions. The phase plane has the coordinates X and $\{dX/dt\}$, and the trajectory on this plane depicts the displacement of X and the velocity $\{dX/dt\}$ as they change with time.

Some fundamental properties of the phase-plane trajectories are readily stated. Through each point in the phase plane, it is possible to sketch a trajectory. Through any one point, only one trajectory may pass. If this were not so, then starting from that point, two different behaviors of the system would be equally possible. Therefore, the system would no longer be determinate but would exhibit a random behavior. For systems under consideration, such behavior is not possible.

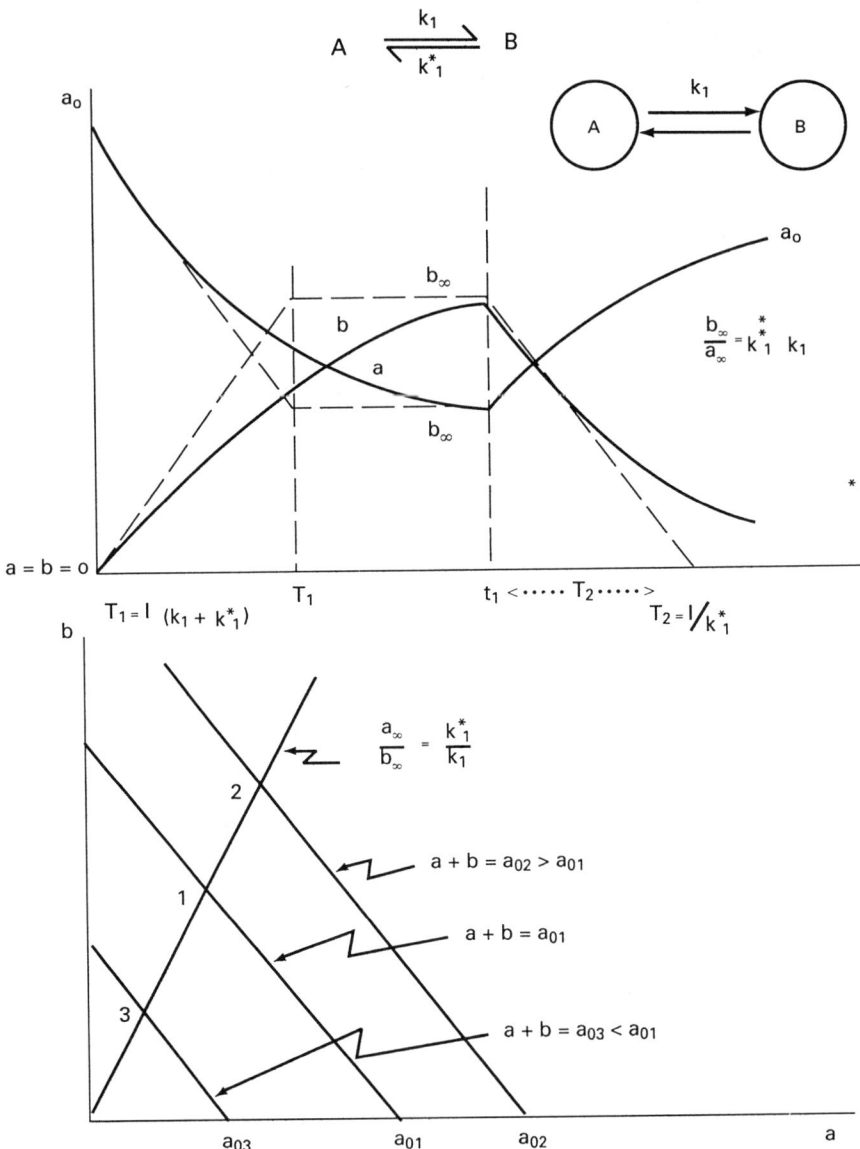

Figure 5.1 Closed chemical system containing a reversible first-order chemical reaction, as described by the two first-order differential equations (5.1) of the text. Plot a: Response to a change in concentration of A from zero to a_0 at time $t = 0$, and at subsequently $t = t_1$, when a reduction in the value of k_1 to zero occurs. Plot b: Phase plane portrait for this reaction, with three different assumed values for a_0. It is assumed that concentrations a and b equal zero, for $t < t_0$. Concentrations of A, a equals a_0 instantaneously at $t > t+_0$, while b equals zero. (Reproduced with permission from Jones, 1973.)

Changes in concentrations are of primary significance in biological processes. The behavior of the system in the phase plane is shown in Figure 5.1b. Any point in the plane represents a specific pair of values of a and b. The lines with negative slope are for the three values of a^o; the line of positive slope gives the ratio of a to b in the stationary state. Their intersection yields the steady-state operating point. Although in the steady state $\{da/dt\} = \{db/dt\} = 0$, this does not imply that the forward and reverse reactions are zero, but rather that they are equal.

OPEN SYSTEMS

Biological processes occur in systems that are open to their environment with which they can exchange matter and energy. Thus, it is possible to identify input and output of molecular species. Laws of equations of continuity and conservation of matter and energy may be appropriately applied. The chemical transformations may be accompanied with the release or storage of energy. Similar to the situation in a chemical reactor, in open systems it is possible to change the concentrations of reactants or the products, as well as the concentrations of catalysts and intermediates. In this manner, different kinds of stationary states are possible to attain.

The reactions of Equation 5.1 can be described again by the set of first-order equations:

$$\{da/dt\} = k_a(a^o - a) - k_1 a + k_1^\star b$$

$$\{db/dt\} = -k_b(b - b^o) + k_1 a - k_1^\star b$$

where

$$a^\star = \{k_1^\star[k_a a^o + k_b b^o] + k_a k_b a^o\}/L$$

$$b^\star = \{k_1[k_a a^o + k_b b^o] + k_a k_b b^o\}/L \qquad (5.3)$$

$$L = [k_a k_b + k_a k_1^\star + k_1 k_b]$$

The variables a^o and b^o are concentrations of A and B in outside solutions of the system which are distinguished from inside concentrations by the boundaries of the system. In Figure 5.2, the two-compartment model example is shown. The reaction specified is a reversible first-order chemical reaction and is described by the two first-order differential equations. Therefore, these exhibit second-order dynamics. Equations 5.3 are obtained on the basis of mass-action law. The dynamic responses given by the differential equations and the final stationary-state values $a^\star$ and $b^\star$ of concentrations of species A and B are presented in Figure 5.2. The stationary-state values are independent of the initial conditions, namely,

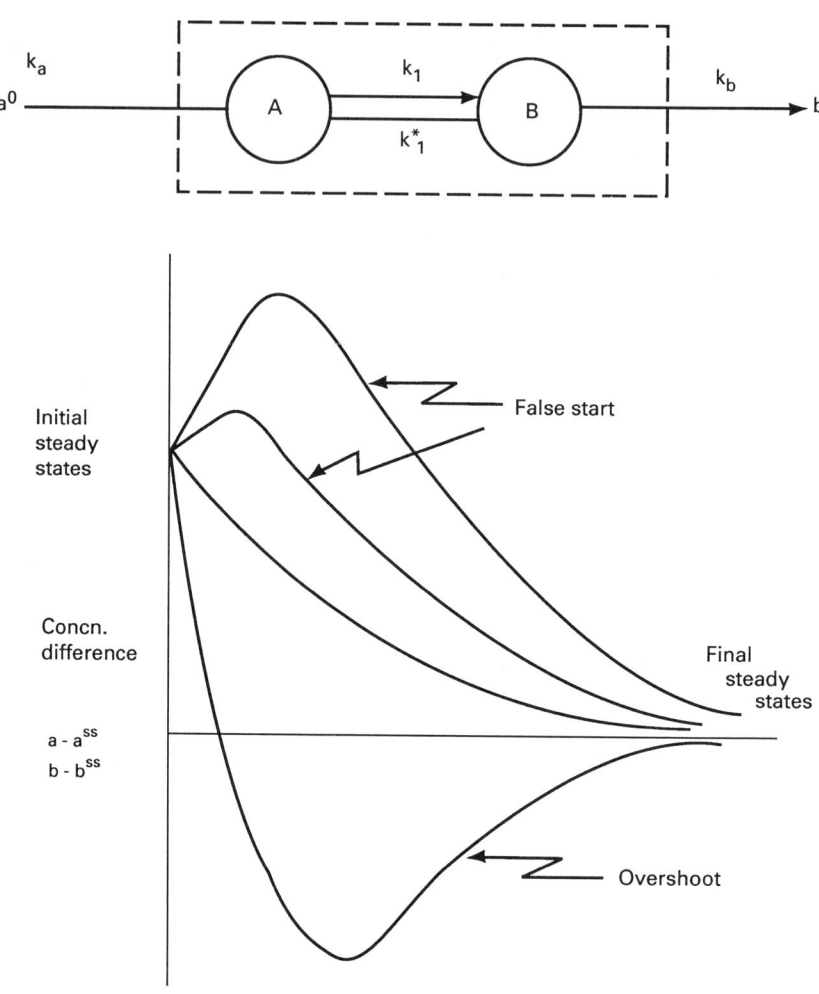

a^{ss} is steady state concentration of a.

Figure 5.2 Open system of two coupled first-order chemical reactions, Equation 5.3, yielding second-order dynamics. System boundaries distinguish between concentrations of species outside, $\{a^0 \text{ and } b^0\}$ from those inside, $\{a \text{ and } b\}$, of the system. The variety of possible dynamic responses are shown in Figure 5.2b. (They all mostly consist of exponential terms. Their form depends on the relative magnitude of the rate constants, k_a, k_b, k_1 and k_1. The concentrations in the final steady state as well as at intermediate time are determined by the rate of addition, k_a, and the rate of removal, k_b. Thus, a control of various concentrations levels are possible in open systems. (Reproduced with permission from Jones, 1973.)

initial concentrations *a* and *b*. This conclusion is an attribute of linear systems.

Thus, for linear systems, the final steady-state values of concentrations depend only on the system parameters and concentrations in the environment. A change in any of the system parameters that produces a new steady state will result in any one of a number of dynamic behaviors presented in Figure 5.2. All of the curves of Figure 5.2 can be shown to be composed of just two exponential terms. The variety of possible forms is brought about by the differences in the two time constants together with their algebraic signs. The values of the time constants depend on the values of the four rate constants. However, the magnitude and sign of each exponential component in the transient depends on the initial conditions.

CONTROL OF REACTIONS BY ENZYMES

Biological processes at the level of the single cell or at the level of multicellular forms of life constitute some of the most intricate problems of chemical kinetics. Signals of cyclic AMP produced by living cells appear to control a number of key vital processes in biology. Since enzymes are very efficient catalysts of most biochemical reactions, very low concentrations (of the order of 10^{-8}–10^{-10} M) suffice to make the reaction proceed at a measurable rate. Only in the presence of enzymes acting as a catalyst, the reactions in biological systems proceed at a rate that has biological significance. The enzymes act as a control signal for the chemical reactions. The kinetic role of enzymes was first given a general formulation by Michaelis and Menten. They proposed that the molecule undergoing the reaction, substrate S, is adsorbed reversibly on a specific site E of the enzyme to form a stable enzyme-substrate complex, whose subsequent decomposition into products is controlling the rate.

When the organism is subjected to a violent change in its environment, it may require new kinds of macromolecules to catalyze the reactions involving new substrates available. In such a situation, the living cell ensures that a new substrate which can itself induce the synthesis of the required macromolecule is available. Such kinds of negative and positive feedback processes and their combinations enable the cells to exhibit a coherent phenomena in the form of an all or none transition or of sustained oscillations.

In biology, often negative feedback processes occur. An example is that in which the product of a chain of reactions reduces the rate of synthesis of a distant precursor of the inhibiting substance. The enzyme reaction, which is a prototype of many similar reactions, can be described as

$$S + E \underset{k_1^*}{\overset{k_1}{\rightleftharpoons}} C \underset{k_2^*}{\overset{k_2}{\rightleftharpoons}} E + P \overset{k_3}{\longrightarrow} P^o \tag{5.4}$$

The reaction represented by Equation 5.4 is the transformation of a substrate substance S into an end product P, under the influence of the catalytic substance E. C denotes the enzyme-substrate complex, which represents the temporary binding of molecule S onto an active site on the enzyme molecule E.

From the law of mass action, one has

$$-\{d[C]/dt\} = 0 = (k_1^\star + k_2)[C] - k_1[E][S] - k_2^\star[E][P]$$

$$-\{d[S]/dt\} = \{d[P]/dt\} = k_1[E][S] - k_1^\star[C] \tag{5.5}$$

The term $k_2^\star[E][P]$ will be absent if the reverse step where enzyme and product collide to produce a complex that is not present.

The assumption that enzyme concentration is much smaller than the substrate concentration and conservation relations requires that

$$[E] + [C] = [E]^{\text{total}}$$

$$[S] + [P] = [S]^{\text{total}}$$

Thus, one obtains

$$-\{d[S]/dt\} = \{d[P]/dt\}$$

$$= \{k_1 k_2[S] - k_1^\star k_2^\star[P][E]^{\text{total}}\}/Z\} \tag{5.6}$$

$$Z = \{k_1[S] + k_2^\star[P] + k_1^\star + k_2\}$$

The concentration of the complex in a steady state can be expressed as

$$[C]_{ss} = [E]^T\{k_1[S] + k_2^\star[P]\}/Z$$

$$[C]_{ss} = \langle k_1[S][E]/\{k_1^\star + k_2\}\rangle \tag{5.7}$$

$$= \langle\{k_1[S][E_o]\}/\{k_1[S] + k_1^\star + k_2\}\rangle$$

In Equations 5.6 and 5.7, it is assumed that the flow of product to the outside described by the first-order rate constant k_3 can be neglected. In the last of Equation 5.7, it is asserted that the reverse reaction of enzyme and product reacting to form the complex does not occur, and thus, the term $k_2^\star$ can be ignored. The ratio $\{[C]_{ss}/[E_o]\}$ can be looked upon as the fraction of occupied sites, in analogy with the Langmuir adsorption concept.

The stationary-state rate of the reaction is given by

$$v = -\{d[S]/dt\} = \langle\{k_1 k_2[S][E_o]\}/\{k_1[S] + k_1^\star + k_2\}\rangle$$

$$= \langle k_2[S][E_o]\}/\{[S] + K_M\}\rangle \tag{5.8}$$

where $K_M = \langle\{k_1^\star + k_3\}/k_1\rangle$ is known as the Michaelis constant. By inversion of both sides of Equation 5.8, one obtains that the plot of the reciprocal of velocity against the reciprocal of the substrate concentration should yield a straight line. The intercept of this line equals $\{k_2[E_o]\}^{-1}$, and the slope equals $\langle K_M / \{k_2[E_o]\} \rangle$. Thus, if the enzyme concentration at time $t = 0$, $[E_o]$, is known, it is possible to evaluate the two constants, K_M and k_2.

Equation 5.6 can be rewritten as

$$-\{d[S]/dt\} = \langle \{V_s/K_M\}[S] - \{V_p/K_p\}[P] \rangle / Y$$

$$Y = \langle 1 + \{[S]/K_M\} + \{[P]/K_p\} \rangle$$

$$V_s = k_2[E]^T \tag{5.9}$$

$$V_p = k_1^{\star}[E]^T$$

$$K_M = (k_1^{\star} + k_2)/k_1$$

$$K_p = (k_1^{\star} + k_2)/k_2^{\star}$$

For initial velocity measurements, the concentration of the product $[P]$ equals zero, so that one can write

$$V_o = v = -\{d[S]/dt\} = V_s/\{1 + (K_M/[S])\} \tag{5.10}$$

The plot of $(1/v)$ versus $(1/[S])$, called the *double reciprocal plot*, yields a straight line. The result of Equation 5.10 is not unique for the single intermediate Michaelis-Menten scheme. This form of the rate equation is obtainable for any number of intermediates in the reaction mechanism. Similar straight lines are also obtained when there is a competitive or noncompetitive inhibitor. In these situations, the value of the Michaelis constant or the intercepts on the y-axis differ (Hammes, 1982).

The application of traditional methods of enzyme kinetic analysis to enzymes compartmentalized by the membrane of the cells, organelles, or vescicular membrane fragments will lead to incorrect estimates of the kinetic constants of the enzymes and incorrect conclusions about the mechanism of reaction (Bunow, 1980). The error is a consequence of concentration differences arising through the reaction processes between the solution outside the cell or organelle membrane and the solution in the interior where the reaction is taking place. Enzymes compartmentalized by the membranes may have restricted access to their substrates due to the limited permeability of those parts of the membrane which separate the enzyme from the substrate present in the solution. A compartmentalized enzyme may be exposed to the concentrations within the compartment, which may differ significantly from those of the external solution.

The appropriateness of the stationary-state assumption is open to question in a system where the amount of enzyme-substrate complex is large, compared to the free enzyme concentration, and $[E_o]$ is not very much smaller than $[S]$.

The stationary-state curves showing $p_0^{\star}$ as the steady-state concentration of the product P to be a linear function of the enzyme concentration e and a saturating function of substrate concentration s are shown in Figure

$$K_m = \{k_2 + k_1^*\} / k_1 = e_\infty \, S / c_\infty$$

$$\dot{p}_\infty^0 = \text{velocity} = k_2 \, c_\infty$$

$$= k_2 \, e_0 \, S / \{K_m + S\}$$

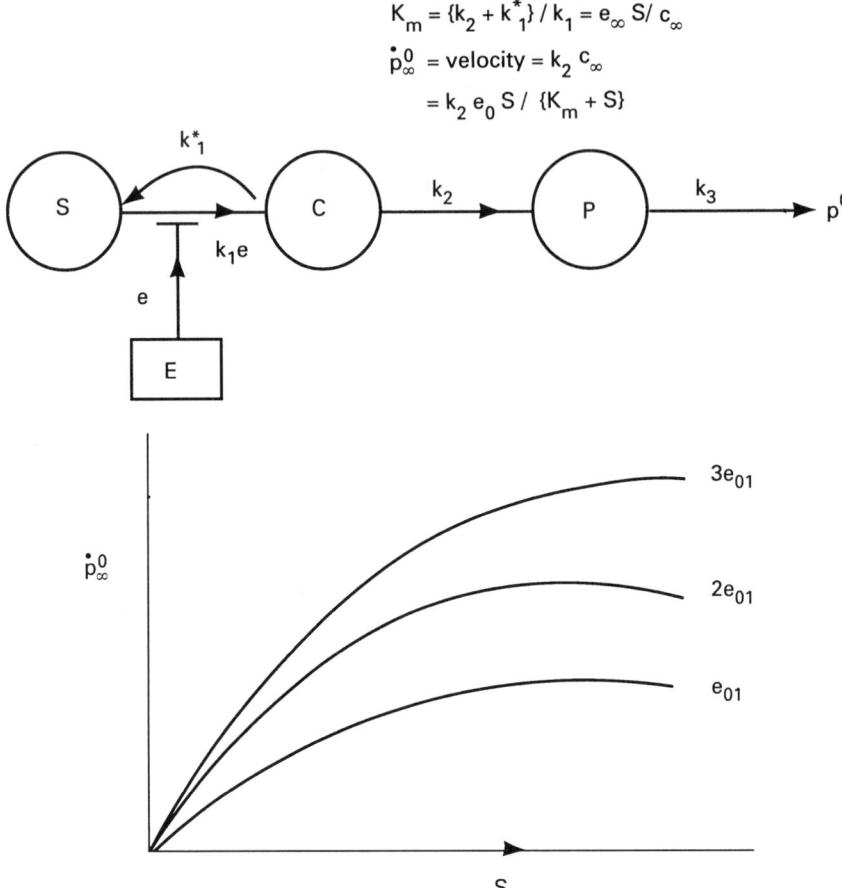

Figure 5.3 (a) Flow diagram for the stoichiometric equations (5.4), with $k_2^* = 0$, for the Michaelis-Menten scheme for enzyme controlled reactions. (b) Steady-state curves showing p_0^* values to be linear functions of enzyme concentrations and a saturating function of the substrate concentrations. (Reproduced with permission from Jones, 1973.)

5.3. K_m is known as the Michaelis constant. The maximum initial velocity V_o is proportional to the total enzyme concentration e^T. The ratio $\{V_s/e^T\}$ is called the *turnover number*, which is a direct measure of the catalytic efficiency of the enzyme.

The Michealis constant K_m is a convenient way of describing the steady-state behavior. The plots of Figure 5.3 may be considered as describing a *control surface*. For small values of substrate concentration s, the rate of production of p_o varies almost linearly with s. However, this reaction saturates for large values of s. Saturation arises since binding sites tend to be more fully filled as substrate concentration increases. At the limit, all binding sites are filled and the maximum value of $p_o^\star$ is set by the turnover number for that enzyme, that is, the maximum number of molecules transformed per unit of time. For each value of e^T, the maximum velocity $\{dp^\star/dt\}$ occurs when all the enzyme molecules are in the complex form C. At saturation, the output $\{dp^\star/dt\}$ becomes a function only of e^T; the biochemical machine is operating at the maximum rate. Since the law of mass action is no longer applicable regarding s, this becomes a zero-order chemical reaction.

The initial velocity V_o at the start of the reaction equals

$$V_o = [V^\star s_o]/[K_m + s_o]$$

$$V^\star = k_2 e^T \tag{5.11}$$

The experimental determination of the Michaelis constant K_m follows, since when $s_o = K_m$, $V_o = [V^\star/2]$. Thus, K_m yields the concentration as the reaction velocity attains half of its maximum value. If this concentration is relatively low, then the reaction is said to be highly specific. The Michaelis–Menten equation can be rearranged into a double reciprocal form, called the Lineweaver–Burk equation:

$$[1/V_o] = [1/V^\star] + \{(K_m/V^\star)/s_0\} \tag{5.12}$$

This yields a straight line for the plot of the reciprocal of the initial velocity against the reciprocal of the initial substrate concentration. The intersection of this line with the vertical axis yields $(1/V^\star)$ and the intersection with the horizontal axis yields $-(1/K_m)$. The maximum velocity $V^\star$ depends obviously on the initial enzyme concentration e_o and on the rate constant k_2. Thus, at high concentrations of the substrate, the speed of the reaction depends only on how many reaction units are there, and on how fast the complex can turn into the product.

Other ways of plotting the initial velocity, for example, the plot of $\{1/V_o\}$ against $\{1/[S]\}$, or the plot of V_o against the ratio $\{V_o/[S]\}$, also yields straight lines, since one can derive that

$$\{1/V_O\} = \{1/V^\star\} + \{K_M/(V^\star[S])\} \tag{5.13a}$$

$$V_o = V^\star - V_o K_M/[S] \tag{5.13b}$$

One should recognize that to assume that the decomposition reaction of the enzyme-substrate complex into the enzyme and its product is an irreversible step is an artificial assumption. On a thermodynamic basis, one can argue that no chemical reaction can be completely irreversible. A reaction may, however, appear as irreversible in an open system where the product is removed fast so that no significant amount of the reverse reaction is permitted to occur. Evidently, the introduction of this irreversible reaction step, as in the scheme of Equation 5.4 denoted by the rate constant k_2^*, will influence the time required to reach equilibrium in a nearly closed system.

The principle underlying the analysis of the kinetic behavior of the reaction of one enzyme with one substrate, resulting in one product, remains the same for more complex systems catalyzed by enzymes, provided the substrate concentration is large compared to the concentration of the enzyme, and the product concentration is essentially zero. A steady–state rate for such a process can be described by an equation of the form

$$V_o = \{dP/dt\} = \{V^\star[S][E]\}/\{K_M^\star + [S]\} \tag{5.14}$$

where $V^\star$ is the maximum velocity at saturating $[S]$ and $K_M^\star$ is the Michaelis constant. The relation between V_o and $[S]$ is described by a rectangular hyperbola. The two constants $K_M^\star$ and $V^\star$ are functions of a number of rate constants. The more complex the mechanisms, the more complex are these functions.

The factors which influence the velocity of reactions catalyzed by enzymes are numerous and varied. It is also a difficult task to obtain the kinetic equations which represent the system from experimental reaction velocity profiles. Substrate concentrations, enzyme concentrations, tempera-ture, pH, and the presence of activators or inhibitors influence the rates. Reaction rates of enzymes are often extremely sensitive to variation of the hydrogen–ion concentration, pH. Enzymes in general exhibit maximum catalytic activity at a definite pH. Two reasons are given for the effect of pH on enzymes. Changes in pH can produce a substantial structural change in the enzyme and this change, which is partial denaturation, can produce a sharp decline in the specific catalytic activity of the enzyme. The second explanation states the effect of pH on enzyme structure is due to ionizations at the active site of the enzyme, with only one of the ionized forms of the enzyme being catalytically active. This kind of behavior can be incorporated kinetically as an additional intermediate mechanism.

The action of the substrate and an inhibitor competing for the same binding site on the enzyme is called *competitive inhibition*. The effect of the competitive inhibitor is to alter the Michaelis constant. One obtains the result, in place of Equation 5.13a,

$$(1/V_o) = (1/V^\star) + \{K_M/(V^\star[S])\}\{1 + [I]/K_I\} \tag{5.15}$$

where $[I]$ is the concentration of the inhibitor and K_I is the dissociation constant of the reaction

$$E + I \rightleftharpoons EI$$

$$K_I = [E][I]/[EI] \tag{5.16}$$

When only the maximum velocity of the reaction is altered by the inhibitor, the inhibition is called *noncompetitive*. This occurs when the inhibitor binds to both the free enzyme and the intermediate with the same equilibrium dissociation constant (Hammes, 1982).

Assume that there are two kinds of substrates, A and B, reacting to yield the products F and G,

$$A + B \rightleftharpoons F + G$$

in which both the substrates react with the enzyme to form complexes, which subsequently decompose into the enzyme and products. This is depicted in the following two substrate reactions with a single enzyme:

$$E + A \underset{k_1^\star}{\overset{k_1}{\rightleftharpoons}} C: K_a$$

$$E + B \underset{k_2^\star}{\overset{k_2}{\rightleftharpoons}} D: K_b$$

$$C + B \underset{k_3^\star}{\overset{k_3}{\rightleftharpoons}} P: K_b^\star \tag{5.17}$$

$$D + A \underset{k_4^\star}{\overset{k_4}{\rightleftharpoons}} P: K_a^\star$$

$$P \rightleftharpoons E + C + D$$

Let e^T denote the total enzyme concentration and K_a, K_b, $K_b^\star$, and $K_a^\star$ denote the dissociation constants of the four reactions stated respectively (Dixon and Webb, 1964). Therefore, at equilibrium one has

$$
\begin{aligned}
K_a c &= a\{e^T - c - d - p\} \\
K_b d &= b\{e^T - c - d - p\} \\
K_b^\star p &= cb \\
K_a^\star p &= ad
\end{aligned}
\tag{5.18}
$$

Under stationary-state conditions, the concentrations of the intermediates c and d do not depend on time. Therefore,

$$K_a K_b^\star = K_b K_a^\star$$

The velocity of the reaction will be proportional to the product concentration p.

$$
\begin{aligned}
V_o &= kp \\
&= \langle ke^T \rangle / \{1 + (K_a^\star/a) + (K_b^\star/b) + (K_a K_b/a)\}
\end{aligned}
\tag{5.19}
$$

The second of Equation 5.19 is obtained by elimination of p from the equations (Haldane, 1930). If the substrate concentration of B is maintained constant, then variation of the concentration of A will yield a curve of the Michaelis type. The apparent affinity for A is observed to depend on the concentration of B. The resultant equation reduces to two simple approximations, when the concentration of B is very large or very small in relation to $K_b^{\star}$, namely,

$$V_o = ke^T/[1 + (K_a^{\star}/a)]$$

or,

$$V_o = \{ke^T b/K_b^{\star}\}/[1 + (K_a/a)] \tag{5.20}$$

respectively. Thus, in the first case, K_M equals $K_a^{\star}$ and in the second case, K_M equals K_a.

If each substrate combines only with its own specific site and there is no effect of one substrate on its affinity for the other, one has $K_a = K_a^{\star}$ and $K_b = K_b^{\star}$. This reduces Equation 5.19 to

$$v = ke[(ab)/(K_a K_b)]/\{[1 + (a/K_a)][1 + (b/K_b)]\} \tag{5.21}$$

Equation 5.21 may be recognized as a product of two Michaelis constants, one for each substrate.

If an enzyme catalyzes a reaction between two indentical molecules combining with the same affinity,

$$Q = \{a/K_a\} = R = \{b/K_b\} \tag{5.22}$$

$$[V_o/(ke^T)] = [R^2/\{1 + R\}^2] \tag{5.23a}$$

or in the reciprocal form

$$[(ke^T)/V_o] = 1 + (2/R) + (1/R^2) \tag{5.23b}$$

Equation 5.23b will give nonlinear reciprocal plots with an upward curvature, similar to those given by a substrate which is also an activator. Various other possibilities for reactions of enzymes with different substrates and mutual interference can be speculated. Reactions between two substances can occur without a definite combination of one of them with the enzyme, the second reaction being involved in a bimolecular reaction with a complex of the first reactant C of Equation 5.17. Thus,

$$E + A \rightleftharpoons C$$

$$C + B \rightleftharpoons E + F + D \tag{5.24}$$

Thus, $V_o = k^\star bc,$ where $k^\star$ is a bimolecular velocity constant. Equation 5.22 is then replaced by

$$\{V_o/ke^T\} = \{Q/[1 + Q]\}R \tag{5.25}$$

Therefore, as the concentration of A is increased, the velocity increases to a limit, but as the concentration of B is increased, it rises linearly.

Substrates can sometimes act as inhibitors or as activators for enzymes. One may also consider the case of the enzyme catalyzing two different reactions simultaneously. It is evident, therefore, that control of biochemical reactions can be accomplished in numerous ways. Many such mechanisms have been proposed and analyzed in various books on enzyme kinetics.

INHIBITION BY LARGE SUBSTRATE CONCENTRATIONS

In most instances, enzyme-controlled reactions operate in a region that is substrate saturated and enzyme limited. That is, the reaction rate is determined almost entirely by the properties and concentrations of the catalyst and is affected very little by changes in the concentration of the substrate molecules. Inasmuch as the substrate in a specific reaction is also the end product of a previous reaction, one would expect minimal changes in the substrate concentrations.

Often one finds that while at lower substrate concentrations the Michaelis equation is obeyed, the velocity of the reaction falls off again at high concentrations. Most enzymes have more than one group, each combining with a particular part of the substrate molecule. In the effective enzyme-substrate complex, one substrate molecule is combined with all these groups, but it is not possible to conceive the formation of an ineffective complex in which a substrate molecule may combine with only one of these groups if the other groups are combined with other molecules. When substrate concentration is high, the formation of ineffective complexes with substrate molecules and active centers increases. This is a case of competitive inhibition by the substrate itself.

When a protein molecule P has multiple binding sites, say n in number to which a ligand molecule can attach, the corresponding sequence of reactions can be represented as

$$L + P_{j-1} \rightleftharpoons P_j; \qquad j = 1, 2, \ldots, n \tag{5.26}$$

where P_j denotes the complex of the protein to which the ligand molecules are attached at j sites, while the remaining $(n - j)$ sites are unattached. It is assumed tacitly that the sites where these j ligand molecules are attached among the n possible sites are all equivalent. When no ligand molecules are attached to the protein, the concentrations of such protein molecules are denoted by p_o, and p_1 similarly denotes the concentration of protein molecules which have one ligand attached per molecule.

The variation of concentrations with time can be represented as

$$\{dp_o/dt\} = -k_1 p_o c + k_1^\star p_1 \qquad (5.27)$$

where $k_1^\star$ denotes the rate constant for the reverse reaction. c denotes the concentration of the ligand molecules L. At equilibrium, there is no time dependence, and one can define an equilibrium *association constant*, K_a,

$$K_a = \{k_1/k_1^\star\} = \{1/K_d\} \qquad (5.28)$$

where K_d is called the equilibrium *dissociation constant*.

Introducing a similar set of association constants K_j for the n reactions, one obtains

$$K_1 = \{p_1/(cp_o)\}$$

$$K_2 = \{p_2/(cp_1)\}$$

$$K_j = \{p_j/(cp_{j-1})\} \qquad (5.29)$$

$$K_n = \{p_n/(cp_{n-1})\}$$

In spite of the fact that the quantities p_o, p_1, and so on, are not determinable experimentally, one can define an average number of r ligands associated with each macromolecule. r is called the *mean association function*.

$$r = \frac{\{\text{total number of ligand molecules combined with } P\}}{\{\text{total number of molecules of kind } P\}} \qquad (5.30)$$

Both the denominator and the numerator of Equation 5.30 are obtainable from experiments. Since there are j ligand molecules attached to each molecule P_j,

$$r = \frac{\{p_1 + 2p_2 + 3p_3 + \ldots + np_n\}}{\{p_o + p_1 + p_2 + \ldots + p_n\}}$$

$$= \frac{\{K_1 c + 2K_1 K_2 c^2 + \ldots + nK_1 K_2 \ldots K_n c^n\}}{\{1 + K_1 c + K_1 K_2 c^2 + \ldots + K_1 K_2 \ldots K_n c^n\}} \qquad (5.31)$$

Equation 5.31 is known as Adair's equation. One can define a saturation function Y as the mean fraction of sites per protein molecule that are occupied, $Y = (r/n)$.

Some simplification occurs when one has identical binding sites. If one assumes that binding at a given site is independent of the state of binding at all other sites (that is, there are only two values of the rate constants, k^+ for the rate of attachment of a ligand to a protein, and $k\star$ for the related reverse reaction) for equilibrium one has

$$0 = -nk^+ p_o c + k\star p_1$$

The factor n appears since there are n possible ways to form the state P_1 from the state P_o, as there are n possible sites for binding. However, there is only one way for the ligand to be removed from state P_1 to form the state P_o.

Similar arguments lead to the conclusion that

$$0 = -(n - 1)k^+ p_1 c + 2k\star p_2$$

since there are only $(n - 1)$ sites available for binding to P_1, and there are two ways to remove the ligand molecule from the state P_2. Thus, the factor 2 is present. One can define the *intrinsic association constant*, K,

$$K = \{k^+/k\star\}$$

$$K_j = (n - j + 1)[K/j]$$

(5.31)

Adair's equation assumes a simple Michaelis–Menten form, with these simplifications:

$$r = \{nKc\}/[1 + Kc]$$ (5.32)

A protein with n binding sites, obeying Equation 5.32 is said to be *noncooperative* or to exhibit *zero cooperativity*.

Hill (1910) assumed that the ligand binding is a completely cooperative process (Edsall and Gutfreund, 1983),

$$P + nL \rightleftharpoons PL_n$$ (5.33)

with no intermediates. The association constant for such a case is

$$K = [PL_n]/\{[P][L]^n\}$$ (5.34)

Using $[P]_t$ for the total protein concentration, the logarithmic expression is

$$\ln K + n \log[L] = \ln\langle[PL_n]/\{[P]_t - [PL_n]\}\rangle$$ (5.35)

Using the notations nH for the Hill coefficient, and $y = \{[PL_n]/[P]_t\}$, one obtains

$$\ln K + nH \ln[L] = \ln\{y/(1 - y)\}$$ (5.36)

Thus, the plot of the fraction y of the bound ligand as a function of the concentration of ligands $[L]$, in terms of the Hill coefficient nH, results in

$$nH = [d \ln\{y/(1 - y)\}/d \ln[L]\}]$$
$$= \{1/\{y(1 - y)\}\}\langle dy/d\{\ln[1]\}\rangle$$ (5.37)

If all sites were identical and independent, the resulting plot would be a straight line with $nH = 1$. With strong cooperative interactions, the value of nH is generally relatively constant over a considerable range of values. In the range where such a simple relation holds, the data can be fitted by an equation with a single constant:

$$y = K[L]^{nH}/\{1 + K[L]^{nH}\} \qquad (5.38)$$

The Scatchard plot, for a system with cooperative interactions, has a positive slope at low ligand concentration values. As $[L]$ increases, the curve passes through a maximum and descends toward zero, as $[L]$ further increases. The slope of the Scatchard plot is

$$\{d(y/[L])/dy\} = (1/[L])\{1 - \langle(y/[L])\{d\ln[L]/d\ln y\}\rangle \qquad (5.39a)$$

while the slope of the Hill plot is

$$\{d\ln(y/(1-y))/d\ln[L]\} = ([L]/y)\{1(1-y)\}\{dy/d[L]\}$$
$$= (1/(1-y))\{d\ln y/d\ln[L]\} \qquad (5.39b)$$

If the Hill coefficient is evaluated at half saturation, one has

$$nH = 4[L]\{dy/d\ln[L]\} = 4\{dy/d\ln[L]\}$$

The maximum of the Scatchard plot for half saturation is

$$\{d\ln y/d\ln[L]\} = 1, \quad \text{for} \quad y = y_{max}$$

The Hill coefficient is thus related to the maximum at half saturation of the Scatchard plot by the relation

$$nH = \{1 - y_{max}\}^{-1}$$

Several problems arise when steady-state enzyme kinetics are used, as described by the Michaelis–Menten scheme, to relate the fraction of the enzyme sites occupied by the substrate. When the Michaelis equilibrium assumption is employed, the equation for the rate v at a specified substrate concentration to the maximum rate V^m, at the saturating substrate concentration is defined by

$$\{V_o/V^m\} = [S]/\{[S] + K_m\} \qquad (5.40)$$

The experimental data may fit an equation of the same form, even when K_m is not a true dissociation constant. The ratio $\{v/V^m\}$ should be used with caution as a measure of fractional saturation (Edsall and Gutfreund, 1983). The binding sites on the biological macromolecules may not be equivalent or independent. Frequently there are different classes of such sites with different association constants for a given ligand. In some cases, the binding sites may be considered equivalent, but there are strong interactions, so that the binding at one site increases or decreases the affinity of other sites for the ligand. Such cooperative interactions can contribute to the sensitive response of receptors involved in many biological systems. The general allosteric model is often not useful in the practical sense, since it has more parameters than Adair's formulation for a macromolecule binding of chemical ligands.

OTHER CONSIDERATIONS

In a sequence of chemical reactions, if one of the reactions is controlled by an enzyme, the rate of conversion of one species into another may or may not obey the law of mass action. When mass-action law is not valid, then this catalyzed step is rate limiting in the steady state. An enzyme-controlled reaction may thus serve as a controlling reaction for an entire sequence of processes if the operation is in the region of substrate saturation. If the system is linear, then the process with the largest time constant would be the controlling one. All other processes with smaller time constants would approach a steady state in a relatively rapid manner, and the one with the largest time constant would in effect set the dynamic response of the whole system.

The dynamic behavior of a single enzyme-catalyzed reaction, as described by the Michaelis–Menten equations, is presented in Figure 5.4. Both nonlinear behavior obtained when the total quantity of the enzyme and substrate is conserved and the linear behavior resulting when the initial enzyme concentration is maintained at a certain level, and the available substrate concentration is also maintained constant, are presented in that figure. Figure 5.4a shows the set of equations governing the reaction when both S and E are present in fixed amounts. This means that neither S nor E are added or lost during the ensuing reaction. When the reaction is initiated, all the substrate is present, but with no synthesis of the product P. Figure 5.4a denotes the course of events as substrate S is depleted and the product P is formed in a closed system. The behavior in Figure 5.4a depicts the concentration of complex c, rising rapidly to a maximum value and then decreasing as P is synthesized. Though the location of the maximum in c and the precise shape of all the curves depend on the relative values of the various rate constants, the ones presented are representative.

In Figure 5.4b, it is assumed that P is removed as it is synthesized according to mass-action law, $\{-k_3 p\}$, and s is maintained at a constant value. The equations are now linear, and the system is described by two first-order linear differential equations. The response of c is a single exponential curve, and that of P is determined by two exponentials. The assumption that S is maintained constant is not too unrealistic if this process

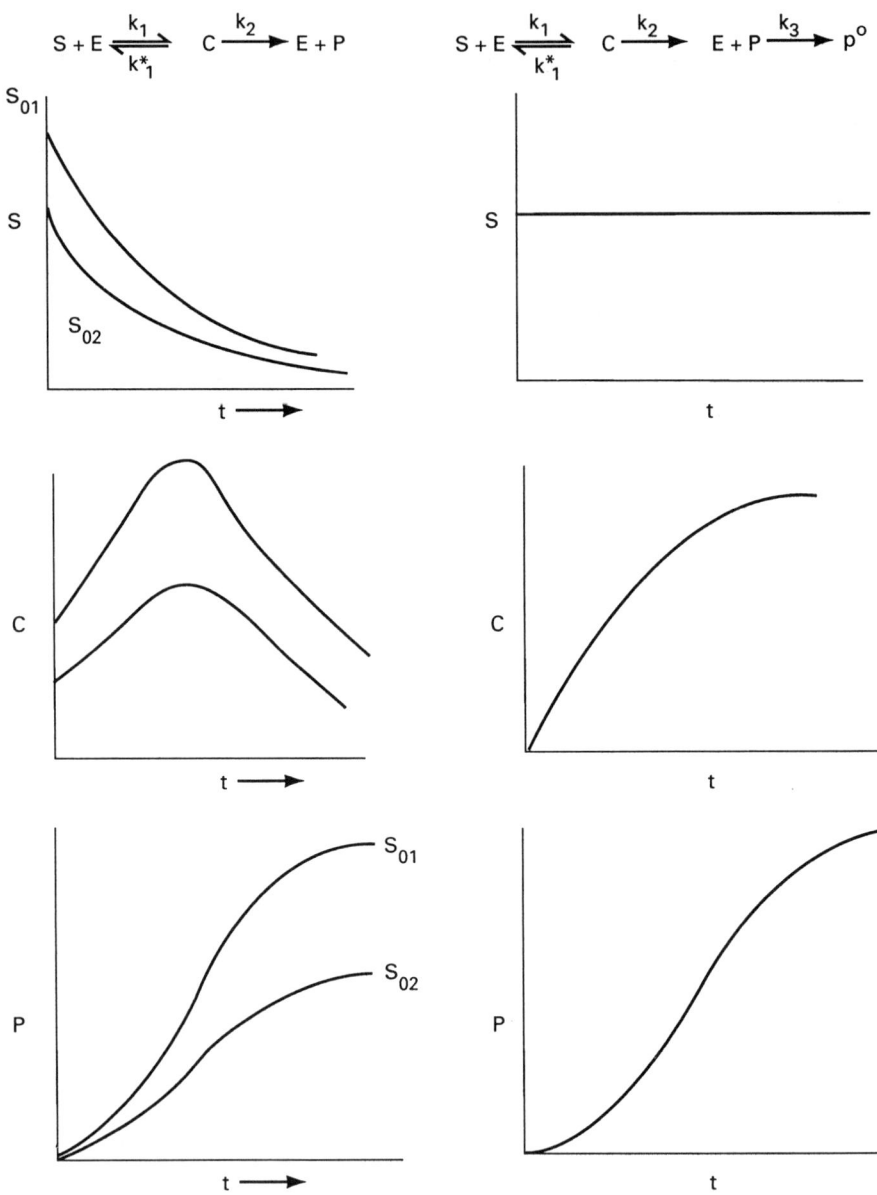

Figure 5.4 The dynamic behavior of a simple enzyme reaction as represented by the Michaelis-Menten equations. (a) Nonlinear behavior when total quantity of enzyme concentration e_0 and substrate concentration s_0 are fixed. (b) Linear behavior when e_0 is fixed and available substrate concentration s is maintained at a constant level, from external supply sources. (Reproduced with permission from Jones, 1973.)

is a step in the synthetic system for which there is an overall feedback control.

FEEDBACK CONTROL BY COMPETITIVE INHIBITION

Competition for the available fixed number of active sites on the enzyme can occur if the enzyme catalyzes the two reactions, namely, conversion of the substrate into the complex, as well as conversion of product P into another substance Q.

$$S + E \underset{k_2}{\overset{k_1}{\rightleftharpoons}} C \overset{k_3}{\rightleftharpoons} P \overset{k_6}{\rightleftharpoons} P^o$$

$$P + E \underset{k_5}{\overset{k_4}{\rightleftharpoons}} Q \tag{5.41}$$

As the concentration of the product p increases, more of the enzyme is utilized in the formation of Q and that available for the reaction with the substrate S decreases.

The steady-state equations corresponding to Equation 5.41 are

$$k_1 e_1 s_1 = (k_2 + k_3)c$$

$$k_3 c + k_5 q = k_4 e_2 p + k_6 p$$

$$k_4 e_2 p = k_5 q$$

$$(k_2 + k_3)c + k_5 q = k_4 e_2 p + k_1 e_1 s \tag{5.42}$$

$$e_2 = (p/s)e_1$$

$$e^T = e_1 + e_2 + c + q$$

where e_1 is the amount of enzyme available for combining with S.

The last of Equations 5.42 shows that the quantity e^T serves the function of a reference input. Since e^T is the total number of active sites on all the enzyme molecules, it will remain a constant as long as the total number of enzyme molecules E is held constant. One may investigate the sensitivity of this system to various kinds of perturbations. With a change in any one of the system parameters, there will be a corresponding change in the characteristic curves in one or the other, or both.

As in any dynamical system, a relationship between the input and output of the metabolic system can be formally described as control characteristics for any given time-independent state. The input is given by the concentration of the substrate or control metabolite, and the output is the rate of change of a concentration of a product of a control unit, or something proportional to it. Control characteristics demonstrate the operational conditions under which the control of an enzyme is effective.

The basis for all states of the metabolic systems and subsystems is the reactivity of the enzymes involved. A classical hyperbolic control characteristic is exhibited by the Michaelis–Menten equation. The sigmoid control characteristics exhibited by a number of systems can be derived from the Michaelis–Menten equation in the following form, analogous to the empirical Hill equation.

$$\log[v/(V^\star - v)] = n \log[x] + \log K \qquad (5.43)$$

where $[x]$ is a metabolite which reacts as a substrate, activator, or inhibitor with an enzyme, $V^\star$ is the maximum reaction velocity, v is any other reaction velocity, K is the apparent Michaelis constant, and n is the number of binding sites for $[x]$ under ideal conditions. When $n > 1$, the Michaelis curve tends to the sigmoid characteristic. This situation is common for the regulatory enzymes and indicates a cooperative binding of a ligand by a protein.

It should be obvious that the control action of a metabolite is zero at the saturation plateau. In the initial portions of the sigmoid curve, where the rate of the enzyme transformation sets the pace of activity, control actions are insensitive. Qualitatively, the gain and damping of control actions can be described by the control ratio, which is the ratio between the two different rates of the system at two given input conditions.

Basically, three principal mechanisms exist to regulate the flow of metabolites through compartments of the cell:

1. The reversible association of existing enzymes with the effector molecules formed during metabolism.
2. Covalent modification of existing enzyme molecules by specific enzyme systems.
3. By induction-repression, which amounts to a change in the amount of the enzyme molecule participating in a particular process. This is achieved by the action of the effector molecules on some site in the enzyme synthesis or degradation mechanisms.

Our current concepts of the regulation of enzyme synthesis are understood on the basis of the operation of lactose operon, controlling the formation of beta-galactosidase and two additional proteins.

One assumes that a repressor R combines with free operators O to form a complex OR, defined by the reaction

$$O + R \rightleftharpoons OR$$

$$K_2 = \{[O][R]/[OR]\} \qquad (5.44)$$

where the square brackets are utilized to denote the concentrations. The total number of operators O^T in the culture population is composed of both free and bound operators.

$$O^T = O + OR$$

One assumes that the repressor combines with n effector molecules E to form a complex RE_n, with greatly reduced affinity to the operator.

$$R + nE \rightleftharpoons RE_n$$

$$K_1 = \{[R][E]^n/[RE_n]\} \tag{5.45}$$

Thus, the two equilibrium relations of Equations 5.40 and 5.41 are assumed to govern the availability of the operator region for initiation by the transcribing machinery. The simplification that forms of the repressor bound to less than n molecules contributes little to the total amount of the repressor, R^T. This results in

$$R^T = R + RE_n \tag{5.46}$$

From Equations 5.36 and 5.37, one obtains

$$R = \{[R^T]K_1\}/\{K_1 + [E]^n\} \tag{5.47}$$

Insertion of this result in Equation 5.44 yields

$$[O]/[OR] = \{K_2(K_1 + [E]^n)\}/\{[R^T]K_1\} = B \tag{5.48}$$

B is the ratio of operons available for transcription to unavailable operons. With no effector present, one obtains

$$Bo = (K_2)/[R^T] \tag{5.49}$$

Bo is the ratio of total repressor concentration to its affinity to the operator region. Thus, one has the equation for induction:

$$\log\{(B - Bo)/Bo\} = n \log E - \log K_1 \tag{5.50}$$

Since B can be evaluated from knowledge of the maximal rate of enzyme production, either by saturation of the system with the effector or from consecutive strains, one can test Equation 5.50 experimentally. A straight line should be obtained for the plot of the left-hand side of Equation 5.50 versus $\log E$. The intercept will yield an accurate measure of Bo. The induction plot yields the three parameters of the induction system.

THE ALLOSTERIC CONCEPT

The term *allosteric* was originally used to describe the effect of an activator or inhibitor introduced at one location in an enzyme on the binding of a substrate at another location. The usage has since changed and the term is now usually employed in connection with the idea that such regulatory effects are due to conformational changes in the macromolecule, introduced by the binding of the ligand. A fairly detailed account of the analysis of allosteric models is presented in Gill et al. (1988). Two classes of allosteric effects were recognized: *homotropic* effects as interactions between identical ligands, and *heterotropic* effects as interactions between different ligands. Chemical ligand-binding reactions play an essential role in the thermodynamic characterization of macromolecular regulation.

The steady-state kinetics of enzyme-catalyzed reactions, according to the Michaelis–Menten scheme, will lead to a hyperbolic curve. However, if the enzyme contains several subunits all of which can bind with the substrate, the shape of these curves will alter from hyperbolic to sigmoid. Sigmoidal curves indicate cooperation in binding. In the case of binding of oxygen to hemoglobin, cooperation implies that at low concentrations the affinity is relatively low, and that it changes to higher values with increasing substrate concentrations. The fact that the curve becomes increasingly steep as the ligand activity increases from zero is the essential feature that marks the curve as cooperative. A curve of this kind can arise only if there are at least two ligand-binding sites. The significant fact about positive cooperativity is the sharp response to a small change in ligand concentration. Such cooperative interactions can make contributions to the sensitive response of receptors involved in many biological systems. The cooperativity of the oxygen binding can be explained by the shift in the equilibria of different structural forms upon oxygenation. The role of conformational changes in the control of enzyme activity became apparent. Evidently, these kinds of binding interactions cannot be described anymore by the simple law of mass action.

Several theories have been proposed to describe this behavior. Adair assumed a set of four different, successively increasing, binding constants. Empirically, any such sigmoidal curves can be adequately described by four such binding constants. However, it could be that the parameters are not related to any actual real phenomenon occurring at the molecular level. In addition, one has to explain why a given binding site changes its affinity upon saturation of other quite distant sites. The general allosteric model has more parameters than the Adair formulation for a macromolecule binding chemical ligands. Thus, it is often not useful in a practical sense.

Consider a macromolecule which has a number of binding sites for each of the several different ligands. Suppose that the macromolecule exists in several different conformations, all in equilibrium with one another, in each of which the affinities for the sites for the various ligands are different. When a given ligand X is added to the system, there will necessarily be a shift of the conformational equilibrium in a direction which will favor the conformations having higher affinities for that ligand. The result is that it will be easier to introduce more of the same ligand than it would be otherwise. The conformational response of the molecule invariably gives rise to positive or cooperative homotropic interactions. Similar considerations apply when the effect of a second ligand Y is considered, except that there are two possibilities here. If the conformations that have a higher affinity for X also have a higher affinity for Y, then the introduction of Y makes it easier to introduce X.

On the other hand, if the conformations with a higher affinity for Y have a lower affinity for X, then the introduction of Y will make it harder to introduce X. The conformational response, in either case, leads to a heterotropic interaction between the two ligands, but in one case it is positive cooperative, and in the other case it is negative anticooperative. The same general principles hold if we broaden the picture to include a change in the state of dissociation or association of the macromolecule in addition to purely conformational changes.

Allosteric enzymes are protein molecules having two or more kinds of active sites. In addition to the normal active site at which the substrate is bound, there are a fixed number of sites, called the *modifier (M)* or the *effector*, at which a small modifier molecule can be bound. The modifier when bound to the enzyme alters the properties of the active site so as to enhance or decrease the catalytic activity. This change within the enzyme molecule brought about by M is termed *allosteric transformation*. Such a transformation is the result of a structural change within the enzyme E. The modifier molecule M is reversibly bound to E, and on the average a certain number of modifier sites will be occupied. The steady-state characteristics are presented in Figure 5.5, where the rate of production of the product P, $\{dp/dt\}$, is plotted against the modifier concentration. For sufficiently large values of m, all the sites are occupied, and $\{dp/dt\}$ becomes almost zero.

Allosteric activation is also possible. Activators and inhibitors have antagonistic effects with regard to active sites. The sigmoidal characteristic curve of an allosteric enzyme suggests a higher-order reaction and a cooperative effect between a number of modifier molecules. This is suggested by the fact that $\{dp/dt\}$ changes more rapidly than m at low values of m.

The term *allosteric* is meant to describe a reaction in which the binding of a ligand molecule to a protein at one site influences the binding of a second ligand at a different site, through a mediation of a conformational change. Since the active and allosteric sites are separate regions on the enzyme molecules, there is no need for any structural similarity between the substrate S and the modifier M molecules. The situation was different in the case of competitive inhibition, where both the molecules P and S competed for the same site. Thus, the allosteric concept opens up many possibilities for control, since the modifier molecule can now control the reaction in

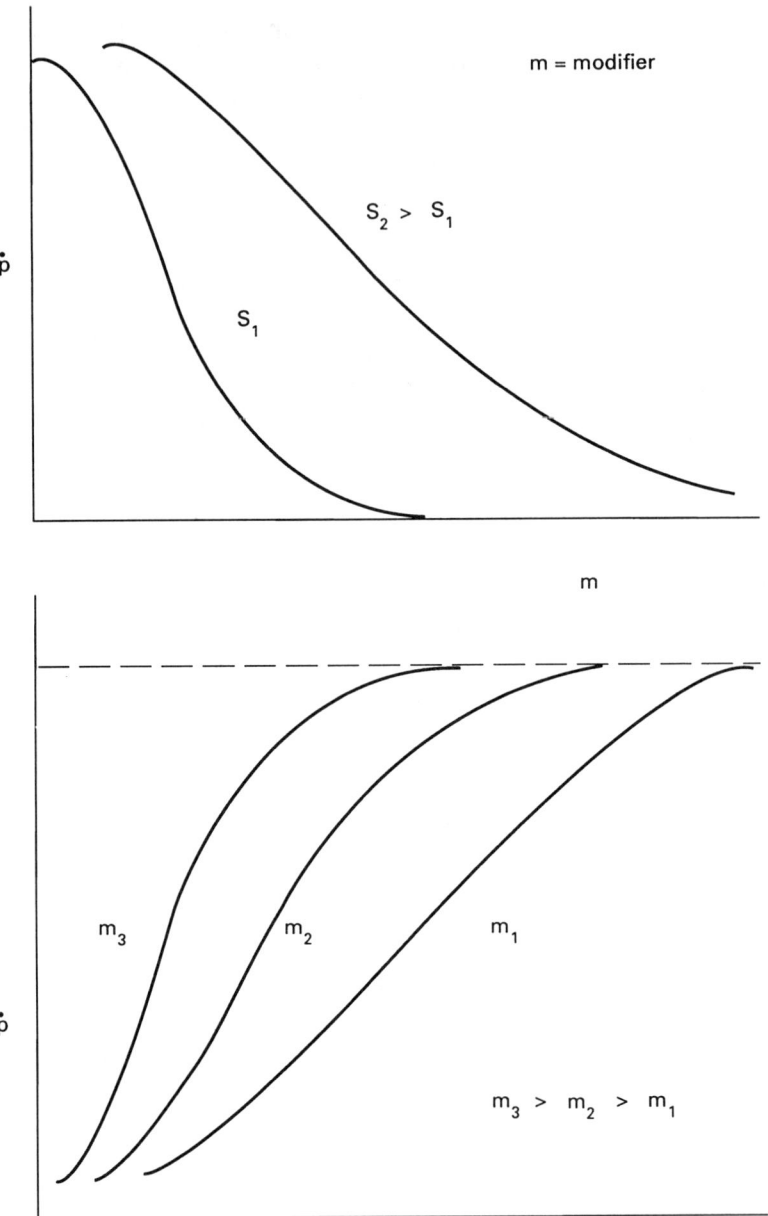

Figure 5.5 Modifier molecule bound to an allosteric enzyme can induce conformational change at the active site, and can produce either inhibiton, shown in Figure 5.5a., or enhancement of the reaction, shown in Figure 5.5b. Note the difference in axes used in (a) and (b). (Reproduced with permission from Jones, 1973.)

which S takes part without the necessity of having any similarity of form. The allosteric molecule acts as a transducer in that the molecular species can now control the reaction involving quite a different kind of molecule. In regulatory enzymes, interactions between spatially distinct active and allosteric sites are manifested by deviations from the Michaelis–Menten kinetics, or by deviations of saturation functions from the hyperbolic law. It is usually believed that cooperative kinetics, that is, the sigmoidal shape of the curve of the initial reaction rate v versus the initial concentration of the substrate $s(o)$ or the allosteric effector $f(o)$ is the pronounced manifestation of the allosteric interactions.

The mathematical description of sigmoidal plots is obtained from Hill's equation, which may be represented for the dependence of v on $s(o)$ $(=s_o)$, the initial concentration of the substrate, as

$$V_o = V\{s^{nH_o}\}/[s^{nH_o} + s^{\star}_{nH_o}]\} \tag{5.51}$$

where V is the maximal rate of enzymatic reaction, V being the limiting value of v at $s(o)$ tending to infinity. $s^{\star}(o)$ is the semisaturation concentration of the substrate at which v equals $(V/2)$, and nH is the Hill coefficient. For sigmoidal plots of v versus $s(o) > 1$, one assumes that the value of nH characterizes a power of interactions between the active sites in the enzyme molecule. Hill's coefficient is calculated on the basis of experimental data as the slope of the plot of $\log\{v/(V - v)\}$ versus $\log s(o)$, in accordance with the equation

$$\log\{v/(V - v)\} = nH \log s(o) - nH \log s^{\star}(o) \tag{5.52}$$

Similarly, the dependence of v on $f(o)$ has the following form:

$$\langle v(o) - v \rangle = [v(o) - v(oo)]\{f(o)^{nH}/[f(o)^{nH} + f^{\star}(o)^{nH}]\} \tag{5.53}$$

Hill's equation of the kind in Equation 5.52 is well fulfilled only if there are strong interactions between the active and allosteric sites. The coefficient nH in this case approaches the number of active sites in the enzyme molecule. If the extreme situation does not take place, Hill's equation is no longer fulfilled, and the value of nH is no longer a constant.

Control of the rate at which enzyme molecules are synthesized becomes a usable control mechanism, since there is a normal turnover. Repression is thus a reduction in the rate of synthesis, so that normal losses of the enzyme will reduce the concentration of these molecules remaining in the cell. Repression is a relatively slow process compared to the inhibition of an allosteric molecule. The rates may differ by several orders of magnitude so that the dynamics of these two processes have quite different time scales. While activation and inhibition provide a fine control of the enzyme-catalyzed reaction, induction and repression provide a relatively crude or coarse control.

Examination of a specific synthetic pathway will usually reveal that more than one control mechanism is used and that these several mechanisms are interconnected in various ways. In a number of pathways, it has been found that repression acts on all intermediate enzymes so that with an excess of the end product all intermediate enzymes are repressed. In feedback theory, coordinate repression would be likened to minor feedback loops from the regulated variable to each of the previous variables in the synthetic pathway. The physiological significance of this arrangement with regard to its effect on the regulated variable should be understood. Inasmuch as repression is a slow process, these minor feedback loops cannot have much to do with the dynamic behavior of the system.

A more reasonable explanation relates to the overall economy of material and energy. By repressing the synthesis of unneeded molecules, an economy in the use of substrates and energy could be accomplished. The induction of enzyme synthesis may act by turning on a structural gene in the DNA molecule. Another possible mechanism is that of deactivating a repressor molecule, so-called *derepression.*

The two modes of enzyme control, the control of enzyme activity and the control of enzyme concentration, by no means represent all possible variations. It is evident that many combinations of these modes are possible. The feedback signals from one molecular species to an enzyme catalyzing an earlier step may also take place. One possibility is the inhibition by an end product in which an inhibitory effect on an enzyme appearing earlier in the synthetic pathway is exerted. A second possibility is that an allosteric molecule exerts control by both its transducer properties as well as by repression of its synthesis. A third form of feedback is one in which a joint feedback from two end products is directed on a single enzyme. This is called a *concerted feedback,* since all end products must be present in excess amounts before there is inhibition of the enzyme activity. A fourth case is called *cooperative feedback,* in which an excess of any of the end products causes a partial inhibition of the enzyme. The simultaneous excess of two or more end products results in greater inhibition than would be predicted from the sum of individual effects. The fifth case, *cumulative feedback,* describes the case in which the inhibitory effects of the several end products act independently.

One may also observe that an enzyme at a certain step can take on a number of different forms, each inhibited by only one of the end products. Although many allosteric enzymes control a reaction that is unique to one end product, an enzyme can catalyze a step preceding a branch point. The implications of such coupling between feedback loops are not well understood.

The basic phenomenon in these situations is the linkage between an equilibrium configuration and the ligand or a set of ligands. Allosteric control effects are bound to arise when a polyfunction macromolecule exists in several different configurations or dissociation states, which are in equilibrium with one another and which show different liganding properties. The association of oxygen with hemoglobin is strongly dependent on pH. As pH is lowered below 7.6, there is a shift toward lower pH of the oxygen saturation as a function of the partial pressure of the oxygen. This shift enables hemoglobin to release more oxygen in the periphery, as carbonic acid builds up in the red cell. It is evident that a linkage exists between oxygen binding and the binding of hydrogen ions to certain sites on the hemoglobin molecule. This reciprocal relation between oxygen binding and hydrogen–ion binding is known as the *Bohr effect.* The origin of this effect is the variations in certain pK values between deoxy-hemoglobin and oxyhemoglobin.

There are also examples in which binding to a system of initially equivalent sites is anticooperative. That is, when one site is occupied, its neighbors become less receptive to the ligand. In the case of binding of NAD+ to glyceraldehyde phosphate dehydrogenase, the enzyme has four equivalent binding sites, one in each of the four subunits. The binding of the first NAD+ molecule to the binding site has an association constant nearly four orders of magnitude greater than the binding of the fourth NAD+ molecule. Thus, chemically, the four subunits of the enzyme are identical, but they are functionally nonidentical. Scatchard plots have the same shape for systems with negative anticooperativity as for those with independent but nonequivalent sites.

The binding of one molecule of a given species modifying the binding affinity for other molecules of the same species is called the *homotropic interactions.* When such binding curves can be greatly altered by changes in the concentrations of other molecules or ions, it is called *heterotropic interactions.* The sigmoid curves for oxygen binding, while retaining their general shape, move progressively farther to the right with increasing partial pressure of carbon dioxide. Therefore, added carbon dioxide diminishes the uptake of oxygen by hemoglobin at a given partial pressure of oxygen. The great biological significance of this relation for the transport of oxygen and carbon dioxide in the blood was recognized immediately. Influx of carbon dioxide from the tissues into the capillary blood will decrease the oxygen-carrying capacity of the blood and release large added amounts of oxygen to the tissues. In the lungs, carbon dioxide is discharged from the blood, so the partial pressure of carbon dioxide falls, and this thereby increases the capacity of the blood to take up oxygen. Thus, carbon dioxide affects the binding of oxygen to the blood, and oxygen affects the uptake of carbon dioxide. Thermodynamically the relation between these two effects is given by the equation

$$\{\partial V / \partial T\}_p = -[\{\partial p / \partial T\}_V / \{\partial p / \partial V\}_T] \qquad (5.54)$$

A considerable amount of progress should be made to master the linguistics of conformational communication and understand the architecture of the energy transfer and energy storage in a macromolecule. How the qualitative properties of the reaction sites are conditioned by their atomic environment is a problem which should be studied using quantum mechanics. Why is it that the oxygen affinity of the heme is so sensitive to conformation of the chain in which it finds itself? Of the several possible roles which heme is capable of playing, why should one be predominant in the peroxidase, another in the myoglobin and still another in the cytochrome?

In spite of its remarkable successes in accounting for a wide range of homotropic and heterotropic phenomena lying at the heart of molecular biology, the allosteric model is still puzzling. From a molecular point of view, the reason for the different conformations having different ligand affinities is not at all clear. Also, why is it that the uptake of a ligand should lead to a conformational change? The introduction of the concept of a symmetry constraint represents an attempt to come to grips with this problem. But assuming this to be correct, it is still not satisfactory. One needs a detailed picture of the conformational changes at the highest level of resolution and an understanding of how these changes are triggered by the ligands.

FREE ENERGY COUPLING

If a metabolic reaction is not at equilibrium, neither the sign nor the magnitude of standard free energy change, G^o, can be utilized to predict the state of the reaction in vivo. Although the value of G^o does not reflect the state of a metabolic reaction in vivo, it is a factor determining the stationary-state concentrations of substrates or products. Thus, in the system (Crabtree and Nicholson, 1988; Hill, 1989)

$$X \underset{v_r}{\overset{v_f}{\rightleftharpoons}} Y \qquad (5.55)$$

where the concentration of Y is determined by the reaction removing Y from the system. This value of Y determines the rate of the reverse reaction. Hence, the concentration of reactant X and the velocity of forward reaction then attains a value such that $\{v_f - v_r\}$ is equal to the steady-state flux through the system. Since G for this reaction should be negative, the minimum concentration of X, for a given flux and $[Y]$, is that which gives a value of G just less than zero, that is, one that just displaces the reaction from equilibrium. Thus, the effect of $[X]$ on G^o can be examined qualitatively by considering its effect on the minimum value of $[X]$, and hence by considering the reaction as if it were at equilibrium. In this case,

$$[X]_{min} = [Y]/K$$

where K is the equilibrium constant, in the direction of $X \rightarrow Y$. This shows that $[X]$ decreases as K increases. Therefore, for a reaction for which $G°$ is negative, in the direction of net flux is able to operate at a lower substrate concentration than a reaction for which $G°$ in this direction is more positive. It is generally advantageous for cells to avoid large concentrations of the intermediate for at least two reasons. A high concentration may be avoided because of the limited solubility or because of significant other side reactions. Second, a high concentration may damage the morphology of the cell.

If the value of $G°$ is too positive to be compatible with a satisfactory concentration of X for the required flux in the direction $X \rightarrow Y$, this can be overcome by kinetically coupling the reaction in parallel with another reaction for which G is large and negative. Thus, in the system,

$$\rightarrow X \rightarrow Y \rightarrow$$

$$\leftarrow A \cap B \rightarrow$$

(5.56)

the reaction $X \rightarrow Y$ is coupled to reaction $A \rightarrow B$ by an enzyme that catalyzes the overall reaction

$$X + A \rightarrow Y + B \tag{5.57}$$

If, at the cellular concentrations of A and B, the value of ΔG for the reaction $A \rightarrow B$ is ΔG_{AB} and if at the cellular concentration of X and Y, the value of G for the reaction $X \rightarrow Y$ is ΔG_{XY} (for the uncoupled situation), the value of ΔG for the coupled reaction is equal to the sum of these component values, since ΔG is a state function,

$$\Delta G_{net} = \Delta G_{XY} + \Delta G_{AB} \tag{5.58}$$

Therefore, if ΔG_{AB} is large and negative, ΔG_{XY} can be positive, provided that the net sum is negative. Thus, there is a net conversion of X into Y, in spite of the fact that ΔG_{XY} is positive. Since ΔG_{XY} is positive, $[X]$ can be smaller than the minimum value in the absence of coupling. This parallel kind of coupling of the reactions helps to prevent a large steady-state concentration of X that would result from an unfavorable value of $\Delta G°$ for the uncoupled reaction. However, this method of offsetting a positive $\Delta G°$ is not possible unless the reactions are kinetically coupled. That is, for ΔG_{XY} being positive and to be offset by a negative ΔG_{AB}, there must be an overall reaction $X + A \rightarrow Y + B$ that is kinetically distinct from reactions $X \rightarrow Y$ and $A \rightarrow B$.

In the absence of kinetic coupling, that is, if the separate reactions $X \rightarrow Y$ and $A \rightarrow B$ merely coexist, there is no interaction between ΔG_{XY} and ΔG_{AB}. Thus, thermodynamic coupling as interactions between free energies is impossible without kinetic coupling. In the example discussed, the large negative free energy change (ΔG_{AB}) did not drive the coupled reaction $(X \rightarrow Y)$ in a given direction. It merely permitted it to proceed in that direction at a physiologically acceptable substrate concentration.

In a sequence of reactions (series–coupled systems),

$$S \rightarrow P \rightarrow Q \rightarrow R \qquad (5.59)$$

one can regard the energy to be transferred from S to R, accompanied by the release of $-\Delta E_{S \rightarrow R}$ kcal/mol of the internal energy of S. The absolute value of the internal energy, however, of one mole of S is unknown, since thermodynamic measurements yield only the changes in the state functions. Thus, the efficiency of the transfer of internal energy from S to R cannot be calculated.

The situation is somewhat different for the case of parallel coupling for the system of chemical reactions,

$$X \rightarrow Y$$
$$A \cap B \qquad (5.60)$$

One has in this situation, from the laws of thermodynamics,

$$\Delta E_{X+A \rightarrow Y+B} = \Delta E_{X \rightarrow Y} + \Delta E_{A \rightarrow B}$$

where the functions in the right-hand side represent the change in internal energy for the individual chemical reactions. If $\Delta E_{X \rightarrow Y}$ is negative and $\Delta E_{A \rightarrow B}$ is positive, the change in internal energy ΔE for the coupled reaction is less than negative than that for the reaction $X \rightarrow Y$ alone. Thus, the latter reaction may be considered to have transferred an amount of chemical energy equal to the difference between $-\Delta E_{X \rightarrow Y}$ and $\Delta E_{X+A \rightarrow Y+B}$, (which from the preceding equations equals $\Delta E_{A \rightarrow B}$) to reaction $A \rightarrow B$. Since the maximum amount of energy that can be transferred from the reaction $X \rightarrow Y$ is equal to $-\Delta E_{X \rightarrow Y}$, one can define an efficiency ε for the transfer, given by the equation

$$\text{Efficiency} = 100 \cdot \{\text{energy transferred/energy available}\}$$
$$= 100 \cdot \{\Delta E_{A \rightarrow B}\}/\{-\Delta E_{X \rightarrow Y}\} \qquad (5.61)$$

Although the conservation relations are dictated for the internal energy by the first law of thermodynamics, one has to be careful in applications of such conservation laws for free energy transfer. The kind of calculations presented are just bookkeeping exercises. They do not in any way make clear how the free energy is transferred from one chemical reaction to another.

CONCLUDING REMARKS

In most instances, enzyme-controlled reactions operate in a region that is substrate saturated and enzyme limited. Thus, the reaction rate is set almost entirely by the properties and concentrations of the catalysts and is affected very little by the changes in substrate concentrations. Inasmuch as the substrate in a specific reaction is also the end product of a previous reaction, one would expect minimal changes in the substrate concentrations. Factors such as changes in temperature, hydrogen-ion concentration, and the presence of small inorganic ions often have marked and persuasive effects on the enzymes.

Although linear systems may be complex, having many components and interconnections, their dynamic behavior is relatively simple in the sense that it can be understood in terms of a few general principles. The dynamic response of even the most complex linear system can be expressed in terms of its modes of free vibration. And although there may be a very large number of such modes, there are only two different types: the decaying exponentials and damped sinusoids.

In contrast, nonlinear systems may exhibit a greater variety of behavior patterns, many of which are far from obvious and not attainable with linear components. With even a single nonlinear component, it is no longer possible to apply the superposition principle, and the response must be calculated in an appropriate manner for each stimulus and disturbance, with few general principles to proceed with.

In linear systems, the behavior is largely independent of the sequence in which the several processes appear in the system. With the introduction of nonlinear components, this sequence of component processes around the loop will have a major influence on the system performance. The presence of a specific nonlinearity and its location in the loop affects the system performance.

In many biological processes, the nonlinear characteristic behavior arises, since what has been previously regarded as a constant parameter is in fact controlled by one of the dependent variables. In the study of any physiological system, one would assume from the beginning that the system contains some nonlinear processes, and that the possibility of finding linear relations must be established by experiments. A linear system, having a single steady-state operating point, will attain that point from any set of initial conditions consistent with the equations of the system and physical constraints. Nonlinear systems frequently having more than one final stable or unstable operating point will follow a trajectory that may be quite different for choices of initial conditions. Change of initial conditions by very small amounts can result in drastically altered final states and behaviors of nonlinear systems.

Biochemical reactions have a marked propensity for oscillations and instability. Cellular metabolism can generate coherent behavior in the form of sustained oscillations and/or spatial patterns extending over distances of macroscopic dimensions. Chemical processes not much different from the ones described in this chapter can exhibit continuing oscillations, and thus may be unstable as a regulating system, or may become an oscillator, causing some biological rhythms.

Unstable limit cycles corresponding to subcritical branches, bifurcating from stationary states in systems with chemical reactions, have been observed. Their existence is due to exponential nonlinearity arising from temperature dependence of the rate constants. From a general point of view, multiple limit cycle behavior has interesting implications. A system at a stable steady state surrounded by two limit cycles is endowed by an intrinsic excitability, in the sense that perturbations from this state exceeding a certain threshold evolve toward a stable periodic region, instead of decaying back to the reference state.

REFERENCES

ATKINSON, D.E., *Science,* vol. 150, (1965), p. 851.

BUNOW, B., *J. Theor. Biology,* vol. 84, (1980), p. 611.

CRABTREE, B., and B.A. NICHOLSON, (1988) in *Biochemical Thermodynamics,* p. 347, ed. M.N. Jones. New York: Elsevier, 1980.

DELISI, C., *Quart. Rev. Biophysics,* vol. 13, pp. 201, 347, in *Biochemical Thermodynamics,* ed. M.N. Jones. New York: Elsevier, 1980.

DENBIGH, K.G., M. HICKS, and F.M. PAGE, *Trans. Far. Soc.,* vol. 44, (1948), p. 479.

DIXON, M., and E.C. WEBB, *Enzymes.* New York: Academic Press, 1964.

EDSALL, J.T., and H. GUTFREUND, *Biothermodynamics.* New York: John Wiley and Sons, 1983.

EIGEN, M., *Quart. Rev. Biophysics,* vol. 1, (1968), p. 1.

GILL, S.J., C.H. ROBERT, and J. WYMAN, in *Biochemical Thermodynamics,* p. 145, ed. M.N. Jones. New York: Elsevier, 1988.

GUTFREUND, H., *An Introduction to the Study of Enzymes.* New York: Academic Press, 1965.

HALDANE, J.B.S., *Enzymes,* p. 83. London: Longmans, 1930.

HAMMES, G.G., *Enzyme Catalysis and Regulation.* New York: Academic Press, 1982.

HESS, B., in *Systems Theory and Biology,* p. 88, ed. M.D. Masarovic. New York: Springer Verlag, 1968.

HILL, A.V., *J. Physiol.,* vol. 40, (1910), p. 190.

HILL, T.L., *Free Energy Transduction and Biochemical Kinetics.* Berlin: Springer-Verlag, 1989.

JONES, R.W., *Principles of Biological Regulation.* New York: Academic Press, 1973.

KOSHLAND, D.E., JR., in *The Enzymes,* vol. IV, ed. P. Boyer. New York: Academic Press, 1970.

MONOD, J., J. WYMAN, and J.P. CHANGEUX, *J. Mol. Biology,* vol. 12, (1965), p. 88.

RENSING, L., and N.I. JAEGER, "Temporal Order," *Proc. Symp. on Oscillations in Heterogeneous Chemical and Biological Systems,* vol. 31. Berlin: Springer-Verlag, 1985.

RUBINOW, S.I., *Introduction to Mathematical Biology.* New York: John Wiley and Sons, 1975.

SEGAL, L.A., *Mathematical Models in Molecular and Cellular Biology.* London: Cambridge University Press, 1980.

chapter 6

Mathematical
Aspects

INTRODUCTORY REMARKS _____

There exist two fundamentally different concepts of physical processes upon which theories of physics are based. According to the first point of view, matter consists of single particles which move in space without undergoing any change. The position of each particle is determinable as a function of time, which is the only independent variable of all processes. This point of view provides the foundation of Newtonian mechanics. The second point of view is based on the field theory of physics. In this, all processes are determined by field quantities which have well-defined values at each point in space. This value is usually a function of time. This one has four independent variables, namely the three Cartesian coordinates and the time.

The differential equations for systems under the influence of external forces are formulated on the basis of Newton's laws of motion and equations of continuity and conservation. From a classical point of view, the behavior of a system of discrete particles is uniquely determined by Newton's laws of motion and the dictates of intermolecular interaction forces acting between the particles. For each particle in the system, one can write three second-order differential equations, which determine the future values of the position and velocity of the particle as a function of time, once the initial values are specified or known.

For a system of N particles, there will therefore exist $3N$ such equations. In principle, all of these equations may be solved and one should find an arbitrary constant of integration for each equation. In order to eliminate these arbitrary integration constants, one should therefore have $6N$ independent pieces of information. These might be the $3N$ position coordinates of each particle at two different times, or the equivalent $3N$ Cartesian components of the velocities. The dynamical behavior of the system is not uniquely defined until there is enough information to determine these $6N$ constants. A broad gap exists between this impossible requirement of a large amount of information and the small number of variables like pressure, temperature, and concentrations of various components required to describe thermodynamically the state of the system.

When two forces of equal magnitude and opposite in directions act on a body and the forces are time independent, the body is in mechanical equilibrium. When two time-independent forces of unequal magnitudes act on a body, the body is subject to a net force equal to the vector sum of the forces and the body suffers a net acceleration, dictated by the relation that *force equals mass times acceleration*. If the forces are time dependent, they could be either monotonically increasing or decreasing with time, or periodically varying with time. The body subject to such forces responds appropriately. If the two forces vary periodically with time and act on the system in opposite directions, the system will undergo an oscillatory motion, whose period and amplitude are determined by the time-dependent behavior of these forces.

The last century of physics can be called the era of *linear physics* with some justification. With few exceptions, the methods of theoretical physics have been dominated by linear equations such as Maxwell's and Schrodinger's, linear mathematical objects (vector spaces), and linear methods such as Fourier transforms, linear response theory, and perturbation methods. The importance of nonlinearity starting with the Navier-Stokes equation and gravitation theory was recognized. However, it was difficult to deal with the nonlinear problems, except as a perturbation to basic solutions of the linearized theory. Specific problems were dealt with almost linear approximations.

Most nonlinear problems defy analytical solutions. One can possibly infer something about the asymptotic long-term behavior of such dynamical systems. It is well known in classical hydrodynamics that a configuration in which a heavy fluid is supported by a light fluid against gravity is unstable. If a state of lower potential energy is available to the system, the system will seek this state, with the extra energy going into the kinetic energy of instability. This thermodynamic approach rests on the assumption that the equations of motion will always allow the system to move in the direction of lower potential energy.

The concept of stability or instability of a system state becomes important in the case of nonlinear problems. One can possibly infer something about the asymptotic long-term behavior of such dynamical systems. Stability analysis depends on many methods, one of which is linearization. For conservative systems, when conservation of energy is valid, one can utilize phase-plane arguments. If an equilibrium state is unstable, then it may be possible to attain stability by means of a feedback control. If the equilibrium of a system is unstable, then any orbit not on its stable manifold moves increasingly towards infinity. An unstable equilibrium forces most orbits to move away from the rest point. If an orbit is repelled by an unstable point, it could tend to a new and stable configuration. Since most problems in biology are nonlinear, some known properties of nonlinear differential equations and a synopsis of their stability properties are presented in this chapter. The properties of solutions of differential equations resulting in oscillations, which are relevant in later discussions of rhythmic behavior, are presented.

OSCILLATIONS

Physical phenomena to the interpretation of which the theory of differential equations has been applied are often of an oscillatory character. Oscillations originate whenever there is an interior force tending to bring a system away from its equilibrium state, and the system responds in such a manner as to bring the system back to its equilibrium state, with the constraint that the system response force is proportional to the displacement of the system from its equilibrium position. In many instances, the observed oscillation is almost periodic. Thus, its mathematical description differs only a little from simple harmonic motion. Closely associated with the problem of oscillations is the problem of stability. Mathematicians have attempted to find methods by means of which the problem of stability could be answered by an implicit study of the defining differential equations.

An important question that should be discussed is how does one determine whether and when a differential equation has a periodic solution?

Periodicity is a very special property, and the answer is difficult to discuss. If $Y(t)$ is a periodic function, it must satisfy the condition

$$Y(t + T) = Y(t)$$

where T is a constant. In addition, all the derivatives of $Y(t)$, if they exist, should also be periodic.

The phenomena associated with oscillatory solutions of differential equations are dependent on the existence of limit cycles in the phase plane. The concepts of both limit cycles and phase planes are presented a little later in this chapter. When such a cycle exists, then the solutions of the differential equations, although they may not be periodic, will approach periodicity. Therefore, the subjects of limit cycles and periodic solutions are related.

Everything which oscillates in a stationary state, in the world around us, is necessarily of the limit-cycle type. It depends only on the parameter of the differential equation itself and not on the initial conditions. Physically, only stable limit cycles are of interest. The unstable limit cycles play the role of separating zones of attraction of the stable cycles in the case when there are many cycles. It is generally difficult to ascertain whether a given differential equation has a limit-cycle solution. Actually aside from well-known equations like Van der Pol, Raleigh, Lienard, and so on, we hardly know anything about the existence of limit cycles for an arbitrary differential equation.

In certain circumstances, one is concerned with the forced oscillations of a nonlinear system in which the forcing function is periodic. If one system possesses a periodic response, a number of the following questions arise naturally.

• Do the neighboring systems also possess periodic responses?
• Is the one periodic response a member of a continuous family of periodic responses?
• Is the stability property of one periodic response reflected in a neighboring periodic response?

Almost all nonlinear problems of the theory of oscillations reduce to the differential equations of the form,

$$\{d^2F/dt^2\} + F(t) + kY[t, F, \{dF/dt\}] = 0 \qquad (6.1)$$

where k is a small parameter, Y is an analytical function in F, and $\{dF/dt\}$ is periodic in time. A typical differential equation for nonlinear oscillation is

$$\{d^2x/dt^2\} + uf[x, \{dx/dt\}] + x = Nwt \qquad (6.2)$$

For N equal to zero, the oscillation is called *autoperiodic*. When w equals an integer, say n, we have *subharmonic resonance*. If the condition that $w = n + @$, where $@$ is a small quantity, one still has subharmonic oscillation, but it is accomplished by synchronization. *Synchronization* consists in the *entrainment* of autoperiodic oscillation by an external drive. If the autoperiodic oscillation exists, the application of a relatively high external drive (*heteroperiodic oscillation*) usually destroys it. This is called *asynchronous quenching*. *Asynchronous excitation* implies the process in which autoperiodic oscillation is released by means of heteroperiodic oscillation. *Parametric excitation* may be defined in the following manner. If a parameter of an oscillatory system is made to vary periodically, with a frequency twice its natural frequency f, the system begins to oscillate with its own frequency. The energy of f oscillation becomes that of $2f$.

HARMONIC OSCILLATOR

For the study of oscillations in biology, it is somewhat educational to review the motion of a simple pendulum, satisfying the differential equation

$$m\{d^2x/dt^2\} = -k^2x \qquad (6.3)$$

where k^2 is a proportionality constant, and m is the mass of the pendulum. The negative sign present in Equation 6.3 indicates that the restoring force acts in a direction opposite to the direction of motion. Considering unit mass and the initial conditions that

$$x(o) = 0, \qquad \{dx(o)/dt\} = 1$$

the solution is

$$x(t) = D \sin(pt); \qquad D = (1/w):$$
$$w = \pm i[k^2/m]^{(1/2)} \qquad (6.4)$$

Similarly, if the initial conditions are

$$x(o) = 1, \qquad \{dx(o)/dt\} = 0$$

the solution becomes

$$x(t) = (1/p) \cos(wt) \qquad (6.5)$$

Both the solutions represent oscillations. The general solution of Equation 6.3 can be written as

$$x(t) = A \exp[iwt] + B \exp[-iwt] \qquad (6.6)$$

where A and B are constants, which may be complex. w equals $\{k/[m]^{(1/2)}\}$. On expanding the exponentials in sines and cosines, one has

$$x(t) = (A + B) \cos(wt) + (A - B)i \sin(wt)$$
$$= C \cos(wt) + D \sin(wt) \qquad (6.7)$$

which may be rewritten as

$$x(t) = E \sin(wt + g) = F \cos(wt + h) \qquad (6.8)$$

A, B, C, D, E, F, g, and h are all constants related to each other by the set of relations

$$E \sin g = C$$

$$E \cos g = D$$

$$F \cos h = C$$

$$F \sin h = -D \qquad (6.9)$$

$$C^2 + D^2 = E^2$$

$$g = \tan^{-1}[C/D]$$

$$h = \tan^{-1}[D/C]$$

E is called the amplitude and g (and h) are the phase angle, which are determined completely.

When friction is present, Equation 6.3 should include an additional term, denoting the resistance that the pendulum mass suffers during its motion, namely, (with $m = 1$),

$$\{d^2x/dt^2\} + 2f\{dx/dt\} + w^2x = 0 \qquad (6.10)$$

The frictional force is proportional to the velocity and f is called the *frictional coefficient*.

Equation 6.3 can be written, with mass equal to unity, equivalently as a pair of first-order differential equations:

$$\{dx/dt\} = p$$

$$\{dp/dt\} = -w^2x \qquad (6.11)$$

The solutions of Equations 6.11 may be written as

$$x(t) = x_o \cos(wt) + p_o \sin(wt)$$

$$p(t) = w\{p_o \cos(wt) + x_o \sin(wt)\} \qquad (6.12)$$

where x_o and p_o are arbitrary constants.

The equations for the trajectories may be expressed as

$$\{dx/dp\} = \{[dx/dt]/[dp/dt]\}$$

$$= -[p/w^2x] \qquad (6.13)$$

where p is the momentum equal to mass times velocity.

The solution of Equation 6.12 is

$$w^2x^2 + p^2 = C^2$$

or,   (6.14)

$$\{x/Cw\}^2 + \{p/C\}^2 = 1$$

The trajectories shown in Figure 6.1 are ellipses with the center at the origin.

When friction is present, Equation 6.3 should include an additional term, denoting the resistance that the mass of the pendulum suffers during its motion. The frictional force is proportional to the velocity, and the proportionality coefficient is called the frictional coefficient, denoted by f. With mass m set equal to unity, one has

$$\{d^2x/dt^2\} + 2f\{dx/dt\} + w^2x = 0$$ (6.15)

One assumes that friction always occurs in a direction opposed to the direction of motion. Equation 6.15 is called the *one-dimensional damped harmonic oscillator problem*.

The auxiliary equation has the roots,

$$[-f \pm \{f^2 - w^2\}]^{(1/2)}$$

so that the general solution of Equation 6.8 can be written as

$$x(t) = A \exp[\{-f + (f^2 - w^2)^{(1/2)}\}t]$$
$$+ B \exp[\{-f - (f^2 - w^2)^{(1/2)}\}t]$$ (6.16)

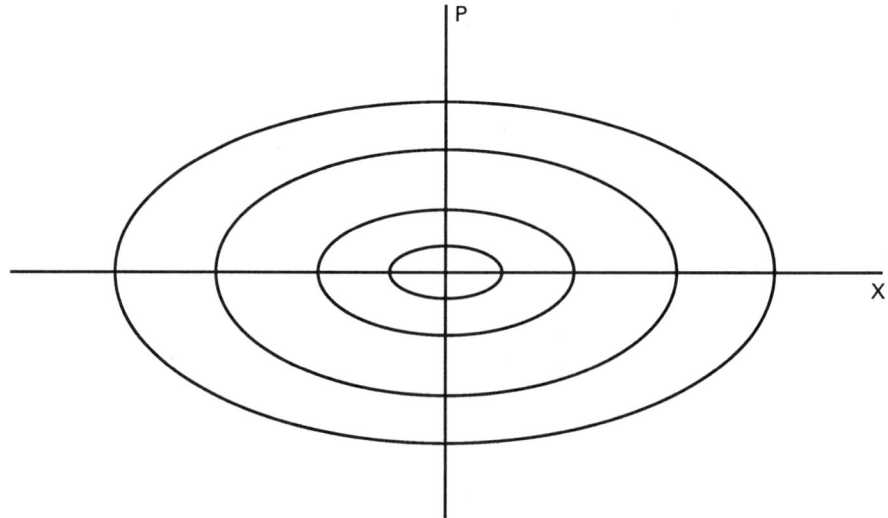

Figure 6.1   Phase-plane diagram for a harmonic oscillator.

With the boundary conditions that at $t = 0$, $x = x_o$, and $\{dx/dt\} = 0$, one may express the solution as

$$x(t) = (x_o/2) \exp\{-ft\}[\{1 + (f/R) \exp[Rt]$$
$$+ \{1 - (f/R)\} \exp[-Rt]] \tag{6.17}$$

where $R^2 = [f^2 - w^2]$. Several special cases which follow are of interest.

1. If $f > w$, R is then real, but smaller than f. Therefore, both terms of Equation 6.17 represent exponential decrease. The motion is, therefore, not oscillatory.

2. If $f < w$, then R is imaginary and may be written as

$$R = iw^\star, \qquad w^{\star 2} = w^2 - f^2.$$

Equation 6.17 may be expressed as

$$x(t) = x_o \exp[-ft]\{\cos(w^\star t) + (f/w^\star) \sin(w^\star t)\}$$

or, equivalently,

$$x(t) = [w/w^\star]x_o \exp[-ft] \sin\{w^\star t + h\}$$
$$h = \tan^{-1}(w^\star/f) \tag{6.18}$$

Equation 6.18 represents a damped sinusoidal motion with the period

$$T = \{2\pi/[w^2 - f^2]\}$$

The amplitude of the oscillations decreases exponentially.

3. If $f = w$, then R equals zero, and $x = x_o \exp[-ft]$. The motion is thus not oscillatory and is said to be critically damped.

Equation 6.15 can again be written as a pair of first-order differential equations:

$$\{dx/dt\} = p$$
$$\{dp/dx\} = -2wx - 2fp; \qquad f > 0 \tag{6.19}$$

For trajectories, $\{dp/dx\} = -[2wx/p] - 2f$.

For small positive values of the frictional coefficient f, the trajectories will spiral into the origin, as shown in Figure 6.2a. Evidently, when $f = 0$, the undamped case is recovered. When f is negative and is of small magnitude, the damped trajectories will spiral away from the origin, as shown in Figure 6.2b.

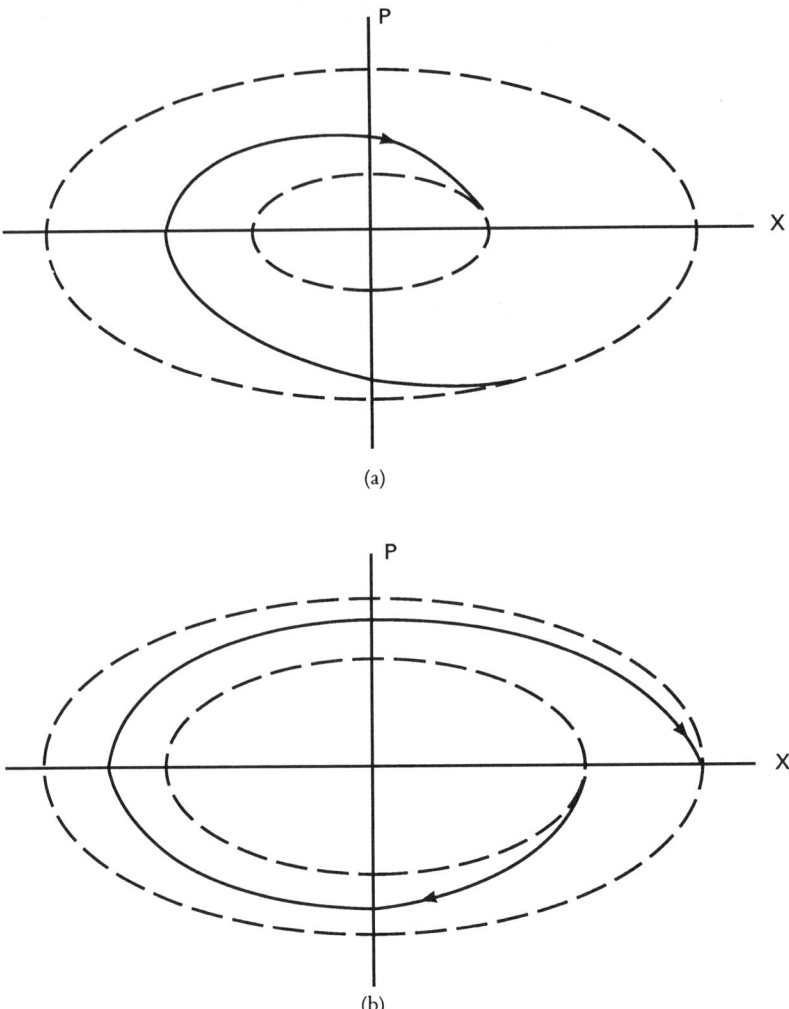

(a)

(b)

Figure 6.2 Phase-plane trajectories of a damped harmonic oscillator.

In biological systems, sometimes one observes *anharmonic oscillations*. Briefly, if one has a term proportional to x^2 in place of the friction term in Equation 6.15, one has an example of an anharmonic oscillator. The differential equation is

$$\{d^2x/dt^2\} + k^2x + Mx^2 = 0 \tag{6.20}$$

The solution can be written as

$$p \, dp = -\{k^2x + Mx^2\} \, dx$$

$$p = \{dx/dt\} = \{N - k^2x^2 - (2/3)Mx^3\}^{(1/2)} \tag{6.21}$$

where N is a constant. Integration of Equation 6.21 leads to an elliptic function.

Consider the linear differential equation

$$\{dx/dt\} + Bx = \cos t \tag{6.22}$$

For $B = 0$, $x = \sin t$ is a periodic solution with a period 2π. The angular frequency w is set equal to unity. The equation for the first variation is

$$\{dy/dt\} + By = 0$$

which for $B = 0$ does not admit a periodic solution of period 2π other than $y = 0$. Thus, there exists a unique family of periodic solutions of period 2π for small magnitude values of B, converging to $x = \sin t$, as B tends to zero. This is the family

$$x = [B/(1 + B^2)] \cos t + \{1 + B^2\}^{-1} \sin t$$

The differential equation when the external force is periodical can be written as

$$m\{d^2x/dt^2\} + f\{dx/dt\} + k^2x = P_o \cos(wt) \tag{6.23}$$

where P_o is the amplitude and w is the frequency of the exterior periodically varying force. One would like to know the amplitude, the frequency, and the phase of the forced oscillation x_1, which is any particular solution without a constant of integration. If f equals zero, one could utilize a function proportional to $\cos(wt)$, which consequently would cancel out. Since f does not equal zero, x_1 would be more complicated. There will be a difference of phase between the force and the resulting oscillation.

Two or more oscillating systems can be coupled. When two particles are interacting such that the first particle is pulled in a certain position of equilibrium by a spring, while the other particle is connected with the first one by another spring, one has a coupled system. Assuming that friction does not exist, and that exterior forces are absent, one can set up the differential equations

$$m_1\{d^2x/dt\} = -C_1x + C_2(y - x)$$

$$m_2\{d^2y/dt^2\} = -C_2(y - x) \tag{6.24}$$

Equations such as Equation 6.24 which are homogeneous can be reduced to a single equation by elimination of one of the unknowns and solutions can be obtained by the method of exponential functions. In real systems, coupling can be obtained by friction or interparticle interactions.

STABILITY CONSIDERATIONS

Suppose that the behavior of a system can be described by two linear first-order homogeneous differential equations,

$$\{du_1/dt\} = a_{11}u_1 + a_{12}u_2$$

$$\{du_2/dt\} = a_{21}u_1 + a_{22}u_2$$

(6.25)

in which a_{ij} are constants. In vector notation, the equations can be compactly written as

$$\dot{U} = \{dU/dt\} = AU \tag{6.26}$$

where the superdot is utilized to indicate the derivative with respect to time. U is a vector with elements u_1 and u_2. If the matrix A, with elements a_{ij}, is nonsingular, that is, it has an inverse, and has eigenvalues N_1 and N_2, the solution of Equations 6.25 can be expressed as

$$u_i(t) = M_1 \exp(N_1 t) + M_2 \exp(N_2 t) \tag{6.27}$$

since there exists a nonsingular matrix P, which diagonalizes the matrix A and uncouples the differential equations.

$$P^{-1}AP = \text{A diagonal eigenvalue matrix}$$

Therefore, $P^{-1}\{dU/dt\} = P^{-1}APP^{-1}U$

$$\{dV/dt\} = \text{Eigenvalue matrix } V$$

$$V = P^{-1}U$$

(6.28)

Equation 6.28 is the uncoupled form of Equation 6.26, and its solutions are simple exponentials. Since the components of the vector V are now known, the components of the vector U are obtained by the operation

$$U = PV$$

If the eigenvalues are real and negative, it is evident that $U(t)$ tends to zero as time t tends to infinity. Thus, the origin is asymptotically stable. On the other hand, if the eigenvalues are positive, then one concludes that $U(t)$ becomes very large as t tends to be large. Thus, the system becomes unstable.

If the eigenvalues are complex, with $N_i = p + iq$, then the solution can be expressed as

$$U(t) = [\exp(pt)]\{R_1 \cos(qt) + R_2 \sin(qt)\} \tag{6.29}$$

in which R_1 and R_2 are suitable vectors. Equation 6.29 states that if p which is the real part of the eigenvalue is negative, then $U(t)$ spirals down to zero from any starting point, thus implying stability. If p is positive, then this describes an unstable situation.

The solution of Equations 6.25 can be written as

$$U(t) = k_1 \exp(N_1 t)C_1 + k_2 \exp(N_2 t)C_2 \tag{6.30}$$

in which C_1 and C_2 are two linearly independent eigenvectors of the matrix A, corresponding to the two eigenvalues.

If p equals zero, then the solution is bounded and this situation is neutrally stable. If only one of the eigenvalues has a negative real part, then the eigenvalues are necessarily real. In this case, $U(o)$ is a multiple of C_1, and $U(t)$ tends to zero as t becomes large.

On the other hand, $U(t)$ tends to infinity if $U(o)$ is initially a multiple of C_2. Therefore, the origin is unstable. A linear combination of these two motions leads to the saddle-shaped orbits that appear to approach the origin and then move away.

PHASE PLANE, SINGULAR POINTS, AND LIMIT CYCLES _____

A differential equation expresses a relation between derivatives and given functions of the variables. It thus establishes a relation between changes in certain quantities and these quantities themselves. This property of a differential equation makes it a natural expression of the principle of casuality, which is the foundation of natural sciences.

Briefly, a differential equation is called *ordinary*, if there is only one independent variable, and *partial* if there are several independent variables. The order of a differential equation is the highest derivative appearing in the equation.

Most of the physical problems that one encounters in biology are nonlinear. Nonlinearity implies that the linear superposition of the two solutions of a second-order differential equation is not a solution of that nonlinear differential equation. Although for some cases of nonlinear equations the solutions may be found by specific transformations of the variables, no systematic method has been discovered for the general solution of nonlinear equations as one has for linear equations. Only qualitative methods ascribed to Poincaré, Liapunov, Bugoliubov, and others are usually utilized. The independent variable that one is concerned with in biological problems is time, since changes with time is the main concern.

There are basically three categories of the stability concept: Laplace, Liapunov, and Poincaré. The first is of such a generality as to be rarely useful, while the second is so restrictive that one is forced to introduce and utilize the third to discuss the stability of periodic solutions.

Laplace stability is a boundedness concept of a very general nature. A system is said to be stable in the sense of Laplace if it exhibits only finite motions; that is, all solutions of the differential equations are bounded as time t tends to infinity. Such a concept of stability is not very useful for quantitative matters in variational problems, since it distinguishes variational effects only as their being finite or infinite.

Liaponov stability is concerned with very stringent restrictions on the motion. It is required that solutions which are once near together, remain near together for all future times, as functions of time. Specifically, it states that a solution $X(t)$ of the system of equations

$$\{dX/dt\} = f(x, t) \tag{6.31}$$

is stable in the sense of Liapunov, if for every $@ > 0$, there exists $\$ > 0$, such that any solution Y of the system of equations, satisfying $[X - Y] < @$, for $t = 0$, also satisfies $[X - Y] < \$$, for all values of $t > 0$.

---

† Let there be a system of differential equations:

$$\{dx/dt\} = f_1(t, x, y)$$
$$\{dy/dt\} = f_2(t, x, y) \tag{1}$$

Let $x = x(t)$ and $y = y(t)$ be the solutions of this system which satisfy the initial conditions,

$$x_{t=0} = x_o$$
$$y_{t=0} = y_o \tag{2}$$

In addition let $x = x^\star(t)$ and $y = y^\star(t)$ be the solutions which satisfy the initial conditions

$$x^\star_{t=0} = x^\star_o$$
$$y^\star_{t=0} = y^\star_o \tag{3}$$

The solutions $x = x(t)$ and $y = y(t)$ that satisfy Equations 1 and the initial conditions in Equations 2 are said to be Liapunov stable as t tends to infinity, if for every arbitrarily small $\varepsilon > 0$, there is a $\delta > 0$, such that for all values of $t > 0$, the following inequalities are satisfied:

$$|x^\star(t) - x(t)| < \varepsilon$$
$$|y^\star(t) - y(t)| < \varepsilon \tag{4}$$

if the equations satisfy the conditions:

$$|x^\star_o - x_o| < \delta$$
$$|y^\star_o - y_o| < \delta \tag{5}$$

Consider the differential equation:

$$\{dy/dt\} + y = 1 \tag{6}$$

Liapunov stability embodies what one desires in the stability concept. However, the time-dependent comparison implied by the inequality

$$[X - Y] < \$^2 \tag{6.32}$$

often precludes stability for certain stationary-state phenomena, which should be considered stable. This difficulty led Poincaré to the concept of orbital stability.

Very frequently one comes across differential equations which have the form exhibited in Equations 6.33. It is possible to eliminate the independent variables between the differential equations

$$\{dx/dt\} = P(x, y)$$
$$\tag{6.33}$$
$$\{dy/dt\} = Q(x, y)$$

and obtain the integral curve

$$\{dy/dx\} = \{[Q(x, y)/[P(x, y)]\} \tag{6.34}$$

Thus, a phase plane (x, y) can be defined. Equation 6.34 is utilized to investigate the behavior of integral curves, or trajectories in that plane. Among the integral curves, those that are closed represent periodic

---

The general solution of Equation 6 is

$$y(t) = Ce^{-t} + 1 \tag{7}$$

It is evident that when $C = 0$, $y = 1$. Find a particular solution that satisfies the initial condition $y_{t=0} = 1$. Find a particular solution which satisfies the initial condition $y^\star_{t=0} = y^\star_o$. The value of C from Equation 7 becomes

$$y^\star_o - 1 = C \tag{8}$$

Substituting this value of C into Equation 7, we obtain

$$y^\star = [y^\star_o - 1]e^{-t} + 1 \tag{9}$$

The solution $y = 1$ is evidently stable. Indeed,

$$y^\star - y = \langle [y^\star_o - 1]e^{-t} + 1 \rangle - 1$$
$$= \{y^\star_o - 1\} \tag{10}$$

tends to zero, when t tends to ∞. Therefore, the inequality in Equation 4 will be fulfilled for an arbitrary ε, if the following inequality holds true:

$$\{y_o - 1\} = \delta < \varepsilon \tag{11}$$

If in Equation 1, the dependence on time t is implicit, then the system of equations is called autonomous.

phenomena. For any autonomous system, time translation invariance is preserved. Thus, to a given closed integral curve C, there corresponds an infinity of motions differing from each other by arbitrary constants. This simplification would not be possible for a nonautonomous system, such as

$$\{dx/dt\} = P(x, y, t)$$

$$\{dy/dt\} = Q(x, y, t)$$

(6.35)

where $\{dy/dx\}$ depends on the variable time t, so that a geometrical representation is not possible.

Assume that Equation 6.35 has a solution written in the form

$$f(x, y) = 0$$

(6.36)

If we have differentiated the first of Equation 6.35, one has three equations in the variables, x and $\{dx/dt\}$. Let us denote this by $y = y(t)$. Analogously, one can obtain a second-order equation in x, which is independent of y and $\{dy/dx\}$. The pair of equations

$$x = x(t)$$

$$y = y(t)$$

(6.37)

are parametric equivalents of the solution given by Equation 6.36. The function $f(x, y) = 0$ is usually called a *function in the phase plane*. This function has an arbitrary constant. Therefore, its graphical representation will be a family of curves in the phase plane called *phase trajectories*.

Consider that Equations 6.37 are valid, as t varies from $-\infty$ to $+\infty$. A curve thus traced will have shapes similar to those three depicted in Figure 6.3. The arrows show the trajectory in which the point $P = (x, y)$ describes the direction. In the first curve (Figure 6.3a), the point P approaches the origin as a limit, as t varies continuously from $-\infty$ to $+\infty$. This limit is called the *focal point* and the motion is characterized as stable.

In the second curve (Figure 6.3b), a different phenomenon is observed. The trajectories are observed to approach asymptotically a fixed curve C. This curve is called the *limit cycle*. One may intuitively infer that the functions of Equation 6.36 will approach the limiting forms which are periodic. Since P approaches the limit cycle as shown by the arrows, the motion is again stable. If, however, the arrows point outward, then the motion would be unstable outside the limit cycle. It would approach the origin as a contracting spiral inside the limit cycle.

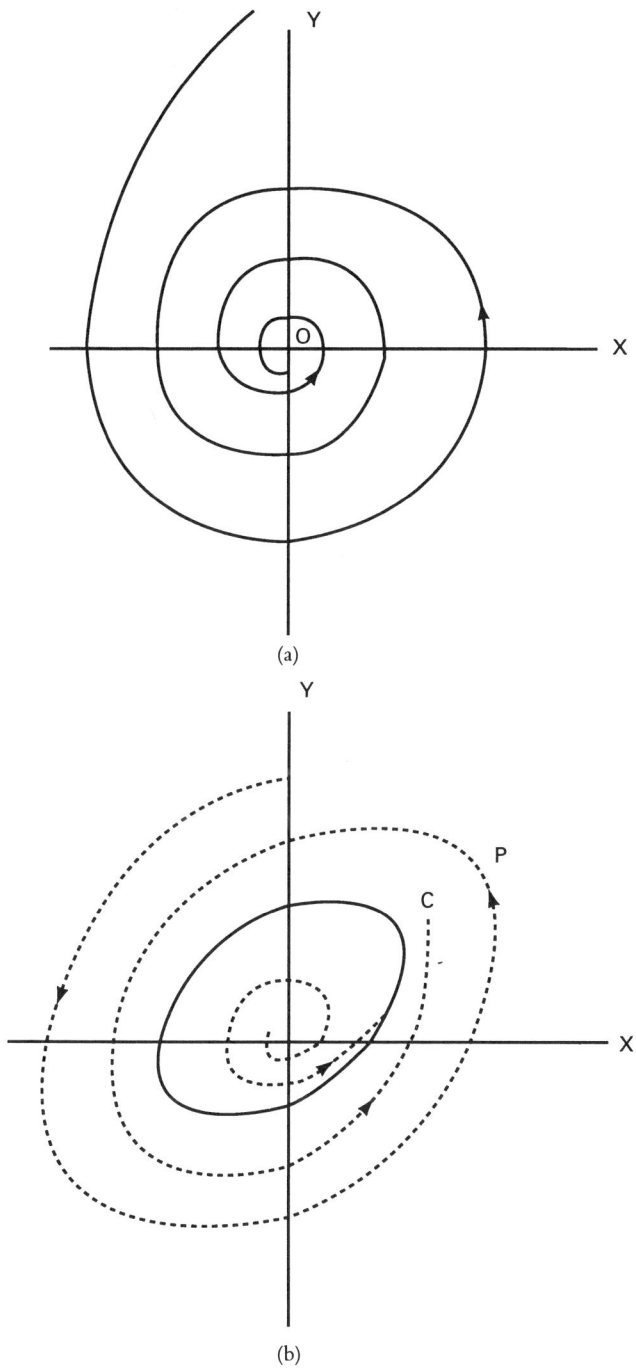

(a)

(b)

Figure 6.3 Plot a: Stable focal point. Plot b: Limit Cycle. Plot c: Saddle point and unstable motion about the origin.

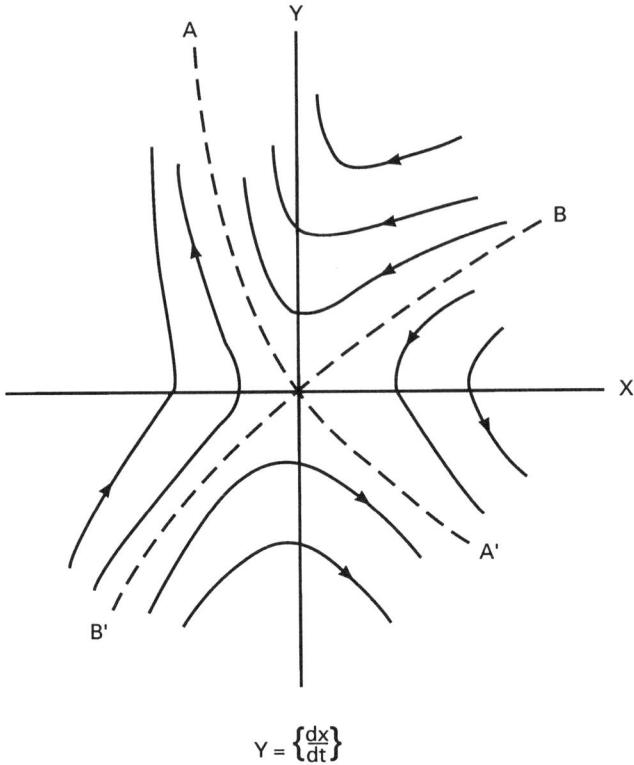

$$Y = \left\{\frac{dx}{dt}\right\}$$

Figure 6.3 *(continued)*

The third curve (Figure 6.3c) represents an unstable motion about the origin, since the trajectories are all hyperbolic curves, which approach the origin but then move away from it. The origin is called a *saddle point*. The dotted lines in this figure, denoted by AA' and BB' toward which the trajectories move asymptotically, separate the plane into regions which contain different motions. These lines are called *separatrix*, for obvious reasons.

We infer that singular points represent the position of equilibrium of the dynamic system, since $\{dx/dt\}$ and $\{dy/dt\}$ approach zero, as the trajectory approaches the singular points. Singular points can be classified as *proper singular point, node, saddle,* and *focus*. Proper singular points are those which all trajectories converge into (in this case, a stable singular point). Saddles are those singular points which some of the trajectories converge and some diverge out. Nodes are those singular points which have an equal number of divergent and convergent trajectories. Focus singular points are those at which the trajectories approach convergently or divergently rotationally.

Relationships exist between limit cycles and singular points. The trajectories reach the limit cycle from the inside. On the other hand, the theory of singular points suggests that when these are unstable, the trajectories leave them. These intuitive considerations suggest that the limit cycles and singular points form certain topological configurations that connect the state of equilibrium (singular point) with the state of stationary motion (the limit cycle). A hydrodynamical analogy is therefore evident. Trajectories act as line of flow and an unstable singular point is a source; a stable limit cycle is thus a sink.

For an unstable limit cycle, the unstable cycle is a source and the stable singular point is a sink. In recent years, the concept of limit cycles has been expanded so as to include cycles that have widespread application in connection with the description of oscillatory phenomena, whose stationary state cannot be described in the phase plane by a trajectory that is nonanalytic. The concept of *nonanalytic limit cycles* has found applications in modern automatic control systems.

Determining stability by the properties of eigenvalues of linear systems leads to the conclusion that the nature of equilibrium is determined only by the linear terms of nonlinear problems. This is somewhat obvious intuitively, since both the function and its first derivative with respect to time decrease indefinitely, so that ultimately only linear terms of first-order magnitude remain.

PHASE-PLANE CONCEPT

There are many good reasons for introducing the phase plane, or the phase-plane concept. Each differential system is expressed as a first-order vector system and the treatment thus gets unified. Pure geometric and vector concepts can be brought to bear anytime as an aid to the understanding of algebraic and analytical processes. In addition, all variables are treated on a par so that there is no artificial distinction made between the initial values of various independent variables. As an example, the initial value of $\{dx/dt\}$ plays the same role in the initial value problem as does the initial value of x.

The plane of the variables $[x, y = \{dx/dt\}]$ is defined as the *phase plane*. The behavior of the integral curves, called the characteristics, in that plane is investigated with the use of Equation 6.34. The motion of the representative point $R(x, y)$ is called the *trajectory*. The integral curves associated with periodic phenomena will have closed trajectories. For autonomous systems, it is possible to replace t by $t + t_o$, where t_o is an arbitrary constant (the phase), and still have the same solution but with a phase shift, t_o. One may assert in this manner that to a given closed integral curve C corresponds to an infinity of trajectories differing from each other by arbitrary constants.

Two definitions may now be introduced:

1. A point (x_o, y_o) for which both $P(x_o, y_o)$ and $Q(x_o, y_o)$ are not zero simultaneously is called an *ordinary point*.
2. A point (x_o, y_o) for which both $P(x_o, y_o)$ and $Q(x_o, y_o)$ vanish simultaneously is called a *singular point*. Singular points represent positions of equilibrium of dynamical systems.

Bifurcation

If for a certain value of a parameter p in the differential equation, the qualitative aspects of the solution remain the same, such values of p are called ordinary values. If, however, for a certain value, $p = p_o$, this qualitative aspect changes, such a special value is called a *bifurcation* or *critical value*. When the bifurcation point is reached, the properties of the solutions of the differential equation change.

There are two kinds of bifurcation points. The bifurcation of the first kind describes the gradual approach of the unstable limit cycles to a stable singular point. In the bifurcation of the second kind, as p approaches p_o, two adjoining limit cycles where one is stable and the other is unstable coalesce (Minorsky, 1962).

LINEAR STABILITY AND FIXED-POINT ANALYSIS

Since the solutions of most differential equations and systems of equations are not expressible in terms of elementary functions, one resorts to approximate methods of integration in the cases of solving concrete differential equations. The drawback of these methods is the fact that they yield only one particular solution. To obtain other particular solutions, one has to carry out the calculations again. Knowing one particular solution does not permit one to draw conclusions about the character of other solutions.

In many problems, it is sometimes important to know not the specific values of a solution for concrete values of the argument, but the type of behavior for changes in the argument. It is sometimes important to know whether the solutions which satisfy the given initial conditions are periodic, or whether they approach some known functions asymptotically. These are questions with which the qualitative theory of differential equations deals. One of the basic problems of the qualitative theory is that of the stability of the solution.

The stability of a particular state of a system is evidently dependent on the regression of the effects caused by external forces in a random arbitrary manner. The stability theory is at the borderline between the deterministic description of the system in terms of the macroscopic equations of motion and the theory of random processes. The very existence of spontaneous fluctuations is a manifestation of the presence of many particles and the interactions among them. When the system is stable, the fluctuations are of no importance and they regress. However, when instabilities of large magnitudes occur due to either intrinsic or extrinsic reasons, fluctuations may amplify and can reach macroscopic level.

The problem of stability is another one of many cases in which the equations of motion of a mechanical system are homogeneous and linear differential equations. If the system is in a motion that does not change with time, this motion satisfies as many conditions of equilibrium as there are degrees of freedom. A rigid body has six degrees of freedom and its stationary motion satisfies six conditions of equilibrium which express the balance between forces and moments. If this equilibrium situation is slightly perturbed, so that all of its forces and moments can be expressed as linear functions of the disturbances, the resulting differential equations are still linear. Thus, problems of stability are described by equations of the same type as equations for oscillations, and can therefore be solved by standard techniques. This method is called the *method of small oscillations*, although in special cases the result may not be periodical.

If the real parts of all eigenvalues are negative, all oscillations are of decreasing amplitude as time increases, and the original motion is stable. If, on the other hand, one or several of the eigenvalues have a positive real part, then the disturbances increase. Any problem of stability can be solved on this basis. However, evaluation of the eigenvalues can be quite cumbersome in many cases.

Consider a set of differential equations:

$$\{dx_i/dt\} = G_i[x_1, \ldots, x_n]; \qquad i = 1, 2, \ldots, n \qquad (6.38)$$

Let the solutions of Equations 6.37 be denoted as $x_1 = q_1(t), \ldots, x_n = q_n(T)$. These solutions are determined uniquely by the initial values of x_i at time $t = t_o$. The $q_i(t)$ describe a trajectory in the Q space. Let us investigate to what extent the trajectories of a particle in Q space depend on the fluctuations in the initial conditions. A trajectory $q(t)$ is denoted as stable when other trajectories $p(t)$, which at time t_o were adjacent to $q(t_o)$, remain adjacent to $q(t)$ at all times $t > t_o$.

Therefore, if at time $t = t_o$, all solutions start in close proximity to $q(t_o)$, the criterion of stability demands that at every subsequent point in time, all trajectories remain close to one another. If as t tends to infinity, this condition is not even intermittently fulfilled, then we call $q(t)$ unstable.

If the two adjacent trajectories $q(t)$ and $p(t)$ come arbitrarily close to each other in the course of time, so that the distance between them vanishes, the trajectory $q(t)$ is called asymptotically stable.

The procedure for determining whether a given trajectory $q(t)$ is asymptotically stable is as follows. If a neighboring trajectory $r(t)$ differs from $q(t)$ by a small quantity $s(t)$, the trajectory $q(t)$ is asymptotically stable, if for large times, $s(t)$ tends to zero. In component form, this implies that

$$r_i(t) = q_i(t) + s_i(t); \qquad i = 1, 2, \ldots, n \qquad (6.39)$$

Substitution of Equation 6.39 in Equation 6.38 yields,

$$\{dq_i/dt\} + \{ds_i/dt\} = G_i[\{q_j(t) + s_j(t)\}]$$

$$i, j = 1, 2, \ldots, n \qquad (6.40)$$

Since all $s_i(t)$ are assumed to be small, the right-hand side of Equation 6.40 can be expanded in a Taylor series, retaining only the leading two terms (linear stability analysis), resulting in

$$G_i[\{q_j^o + s_j\}] = G_i[q_j^o] + \{\partial G_i/\partial s_k)s_k$$

$$\{dq_i^o/dt\} + ds_i/dt\} = G_i[q_i^o] + \{\partial G_i/\partial s_k\}s_k$$

from which it follows that

$$\{ds_j/dt\} = \{\partial G_j/\partial s_k\}s_k(t) \qquad (6.41)$$

In general, it is difficult to solve the set of differential equations from Equation 6.41, since both q_j and $\{dG/ds\}$ are time dependent. When all q_j's are constants, one has

$$\{\partial G_i/\partial s_k\} = \{\partial G_i/\partial x_k\} \qquad (6.42)$$

and one can define

$$a_{ik} = \{\partial G_i/\partial x_k\} \qquad (6.43)$$

The set of first-order differential equations with constant coefficients can be written in a matrix form

$$\{dS/dt\} = AS \qquad (6.44)$$

where A is the matrix with coefficients, a_{ik}.

The eigenvalues of A are obtained by solving the characteristic equation. If the real parts of all eigenvalues are negative, all perturbations decay exponentially with time, so that the solution is asymptotically stable. As soon as one eigenvalue with a positive real part exists, one can always find an initial trajectory which deviates from $q(t)$, so that the solution is unstable. The situations where no eigenvalue is positive, but some can be zero, exhibit *marginal stability*.

The differential equations are usually very complicated and are solvable only by numerical methods. In many cases, the interest in these equations is in their long-term behavior. A suitable procedure for obtaining this information is the *fixed-point analysis*.

Fixed points are states of a dynamic system that are invariant with time. The fixed points are those points in, say, an N-dimensional concentration space, for which the composition of the model system no longer changes. This implies that the positions of all fixed points are determined by the stationary-state conditions:

$$\{dX/dt\} = 0 \qquad (6.45)$$

The macroscopic changes in the dynamic behavior of the system for small perturbations in concentrations at given fixed points should be investigated.

The linear stability analysis will lead to a description of the dynamics of the perturbed system in terms of a spectrum of normal modes with respect to time constants w_k: the latter appear as eigenvalues of a set of differential equations linearized around a fixed point.

The basic principle of fixed-point analysis is as follows: The dynamics of the system are determined by the form of potential surface in the concentration space. This potential surface possesses in general a set of local maxima and local minima which correspond to fixed points of the system. The set of fixed points thus presents an abstraction of the most important structural features of the complete potential surface. Finding the lines (valleys) joining individual points enables one to make general statements about the dynamics of the system, without describing the entire potential surface. The time development of the system will follow downhill, as a rule.

There are four classes of fixed points:

1. The stable fixed points or sinks: All eigenvalues have negative real parts, so that all perturbations decay exponentially. The system reacts to fluctuations with an automatic compensation for the fluctuation.

2. The sources: All eigenvalues are real and positive, which means that all directions of fluctuations are unstable. Whenever a perturbation appears, the system tends to reinforce it. The corresponding fixed point has the property of a source in that the system tends to move away from it.

3. Saddle points: One eigenvalue is real and positive and others are real and negative, so that one direction of fluctuation is unstable. Saddle points thus behave as sources, but only in particular directions.

4. General unstable fixed points: Some eigenvalues are zero or pure imaginary. Systems with pure imaginary values make up so-called centers, whose trajectories consist of a family of concentric circles.

NATURAL OSCILLATIONS IN AN ELECTRICAL CIRCUIT _____

In a circuit containing resistance R, capacitance C, and inductance L, the sum of partial electromotive forces equals the external electromotive force. For natural oscillations, when the external electromotive force equals zero, one has the differential equation

$$L\{dI/dt\} + RI + (q/C) = 0 \qquad (6.46)$$

Or, since $I = \{dq/dt\}$,

$$\{d^2q/dt^2\} + [R/L]\{dq/dt\} + [1/LC]q = 0 \qquad (6.47)$$

This equation is of the form of Equation 6.50 and the constants are

$$b = [R/2L]; \qquad w = (LC)^{-(1/2)} \qquad (6.48)$$

If oscillations are to occur, $w > b$; that is, $R < 2[(L/C)]^{1/2}$. One has

$$q = \{1 - [R^2C/4L]^{-(1/2)}\}q_o[\exp\{-(R/2L)t\}] \sin H$$

$$H = Mt + \delta$$

$$M = \{(LC)^{-1} - (R/2C)^2\}^{(1/2)} \qquad (6.49)$$

$$\delta = \tan^{-1}\{(4L/CR^2) - 1\}^{1/2}$$

The initial conditions are that at time $t = 0$, there is no current and the condenser has a charge q_o.

FORCED OSCILLATIONS _____

In recent years, considerable interest has been evinced on the influence of electromagnetic waves and oscillating electric fields on biological systems. It is educational to consider the forced oscillations of the system of (charged) particles in this respect.

The damped harmonic oscillator is characterized by the presence of an additional force, proportional to the velocity of the particle, at the specified instant. This is known as the *frictional force*. The equation of motion for this system is given by the differential equation

$$\{d^2x/dt^2\} = -w^2x - 2F\{dx/dt\} \qquad (6.15)$$

where F is called the frictional coefficient and is regarded as a constant. If F is positive definite, the force arising from the friction is also directed opposite to the displacement from the equilibrium position. The solution of Equation 6.15 has been discussed previously.

For forced oscillations, Equation 6.46 with an additional function which is time dependent, exists in the right-hand side. This function which expresses the impressed force divided by the mass of the oscillating system is a sinusoidal function of the time in most applications. When this additional function is of the form $f_o \sin(at)$,

$$\{d^2y/dt^2\} + 2F\{dy/dt\} + w^2y = f_o \sin(at)$$

the particular integral can be written as

$$\text{Particular integral} = [\{\exp[-b + R]t\}/2R] \int \{\exp[b - R]t\}f_o \sin(at)\, dt$$

$$- [\{\exp[-(b + R)]t\}/2R] \int \{\exp[b + R]t\}f_o \sin(at)\, dt \quad (6.50)$$

where $R = \{b^2 - w^2\}^{(1/2)}$

The complete solution of the forced oscillation problem is obtained by the addition of this particular integral to the solution of Equation (6.15).

$$x(t) = K\{(w^2 - a^2) \sin(at) - 2ba \cos(at)\}$$
$$+ e^{-bt}[C_1 e^{Rt} + C_2 e^{-Rt}] \quad (6.51a)$$

The second term, namely, the complementary function decays exponentially with time and will damp out.

One obtains the result that the amplitude of the oscillations

$$K = \{f_o \langle [w^2 - a^2]^2 + 4(ab)^2 \rangle\} \quad (6.51b)$$

has a maximum, when the impressed (angular) frequency has the value

$$a = \{[w^2 - 2b^2]^{(1/2)} \quad (6.51c)$$

This is identified as the condition of resonance between the impressed force oscillation and the oscillating system. If b equals zero, there occurs a resonance catastrophy, since in this case, the amplitude is infinity, when a equals w.

For electrical systems, with an impressed electromotive force, $E_o \sin(at)$, $(f_o = \{E_o/L\})$ such a resonance occurs when,

$$a = [(1/LC) - (R^2/2L^2)]^{(/12)} \quad (6.52)$$

If a lattice is in equilibrium, the corresponding energy density is a minimum. A structure may be stable with respect to homogeneous deformations, and yet be unstable for other types of small deformations. If it is found by solving the equations of motion that the frequencies of all the normal modes are real, then the lattice is stable for all small deformations; otherwise it is unstable.

An equation of the form of Equation 6.15 with the right-hand side being a periodic function also describes the response of ordinary matter to an impinging electromagnetic wave. These waves are polarized in such a manner that the electric vector is along, say, the y-direction, when incident upon a charged particle, including those present in biological systems, will exert a force equal to

$$eE_o \sin(at)$$

upon a particle with charge e. E_o is the amplitude of the electric vector of the electromagnetic radiation and

$$f_o = (e/m)E_o \tag{6.53}$$

where m is the mass of the charged particle, with charge e. The solution is given by Equation 6.51. x represents the displacement of the charged particle at time t. This gives rise to a dipole moment (ex). a is the frequency of the monochromatic radiation. Polarization equals the dipole moment per unit volume of the material and is obtained by multiplying the dipole moment due to one electron by the number of displaceable electrons per unit volume.

The influence of electromagnetic fields on a system of charged particles is governed by Maxwell's equations:

$$\nabla \times E = -\{1/c\}\{\partial B/\partial t\} \tag{6.54a}$$

$$\nabla \times B = (4\pi/c)I + (1/c)\{\partial E/\partial t\} \tag{6.54b}$$

$$\nabla \cdot E = 4\pi\rho \tag{6.54c}$$

$$\nabla \cdot B = 0 \tag{6.54d}$$

where E and B denote the electric and magnetic fields. ρ and I denote the charge and current densities, respectively, and c denotes the velocity of light. The charge density ρ and current density I are connected by the equation of continuity, readily obtained from Maxwell's equations:

$$\nabla \cdot J = \{\partial \rho/\partial t\} = 0 \tag{6.55}$$

J is current density and ρ is charge density.

If Equations 6.54a, 6.54b, and 6.55 are considered as equations of motion for E, B, and ρ, and consider Equations 6.54c and 6.54d as initial conditions, one obtains

$$\nabla \cdot [\nabla \times E] = -(1/c)\{\partial/\partial t\}[\nabla \cdot B] = 0 \tag{6.56}$$

and

$$\nabla \cdot [\nabla \times B] = (1/c)\{\partial/\partial t\}[\nabla \cdot E - 4\pi\rho] = 0 \tag{6.57}$$

This shows that the initial conditions in Equations 6.54c and 6.54d are valid all the time, if valid initially. Thus, the validity of the Poisson equation is independent of the state of the system.

For a system of charged particles, the charge density and current densities are defined as

$$\rho = e \sum_k Z_k C_k$$

$$I = e \sum_k Z_k C_k V_k \qquad (6.58)$$

where Z_k is the signed valence charge number and C_k is the concentration of ions of kind k. e is the magnitude of the charge of an electron. V_k is similarly the velocity of ions of kind k. For a system of electrolytes consisting of massive ions, the ratio of velocity of charged particles over the speed of light is a negligible quantity. Therefore, the terms involving $(1/c)$ in the preceding equations can be neglected. The stability considerations and criteria for oscillations developed in connection with plasma physics (Chandrasekhar, 1960) can be utilized. The equation of motion for charged particles can be written as

$$m\{dV_k/dt\} = eEZ_k \qquad (6.59)$$

For the description of the vibratory motions, a set of coordinates, q_i, which are linear functions of the displacements of the particles from their equilibrium positions, can be introduced. The particles may be assumed to move independently of one another, each as a sinusoidal function of time

$$A_i \sin\{2\pi n_i t + d_i\}$$

where n_i is the frequency of the ith particle, determined by the nature of forces acting upon it, and A and d are arbitrary constants. For macroscopic systems, the number of such vibration frequencies is large. Certain assumptions regarding the distribution of frequencies are invoked in the theories of specific heats of Einstein and Debye. The motions of the particles in an elastic continuum can be resolved into elastic waves of the form (Born and Huang, 1954):

$$U(X, t) = AN \sin\{2\pi Y \cdot X - 2\pi nt + d\} \qquad (6.60)$$

$U(X, t)$ is the displacement of the medium at a point x at time t. Equation 6.60 describes a plane wave of frequency n with the displacements in the direction of the unit vector N, with the wave normal parallel to Y and the wave number equal to Y. Y is called the *wave number vector*. A and d are the usual amplitude and phase factor associated with an oscillatory motion. For any given Y, there are three independent modes of elastic waves. Due to the presence of the arbitrary phase, d, a similar cosine wave does not constitute any further independent mode of vibration. Equation 6.60 represents the solutions of elastic equations of motion and need not constitute the normal mode of vibrations. The normal vibrations are linear combinations of the preceding solutions of the same frequency, which satisfy the boundary conditions on the surface.

When a system consists of two different kinds of particles, with their own equations of motion, the solution leads to a sinusoidal wave for the displacements of either kind of particle. The two waves describing the displacements of the two types of particles, which differ in mass, have the same wavelength and frequency. In addition they share the same arbitrary phase and amplitude factors. The imaginary part of the complex solution does not give additional independent solutions.

The frequencies in general fall into two distinct branches, corresponding to two alternative solutions called the *acoustic* and *optical branches*, corresponding to longitudinal and transverse vibrations. For long wavelengths, the frequencies are linearly proportional to the wave number. The long acoustic waves are essentially determined by the elastic properties of the system: The stability against homogeneous deformations ensures that the frequencies of the long lattice waves will be real. The long waves of the acoustic branch are identical with the longitudinal elastic vibrations.

The optical vibrations, on the other hand, approach finite frequency. In the long optical vibrations, the opposed motions of oppositely charged particles in the system give rise to a net oscillating dipole moment of the cell, which is absent in the long waves of the acoustic branch. In the acoustic case, the oppositely charged particles move in unison.

It is evident from these considerations that in a system consisting of various kinds of particles with different masses and charges, many possibilities of vibrations under the influence of an external oscillating field exist. Biological systems have numerous kinds of charged particles with different masses. The influence of electromagnetic waves on biological systems has its origin in these kinds of responses. Many rhythmic phenomena can be accounted for by consideration of the effects caused by such different forces, resulting in the coupling of various effects.

VAN DER POL OSCILLATOR

The frictional coefficient f introduced in the case of the damped oscillator measures the degree of damping of the system, depending on its magnitude. Thus, it can be called another system parameter. If f is taken also to be a function of time, instead of being a constant, additional interesting results arise. As before, f can now become functions of the state parameters, in which case one has the situation of a feedback system. Evidently, the system behavior is dependent on the time dependence as well as the state variable dependence of the forcing function f imposed on the system.

The results of a particular choice of

$$f(t) = f[x(t)] = -@\{1 - x^2\} \qquad (6.61)$$

where @ is a small positive number can be considered.

With this choice of f, the equations describing the harmonic oscillator become

$$\{dx/dt\} = p$$
$$\{dp/dt\} = -k^2x + @\{1 - x^2\}p \qquad (6.62)$$

The equation of motion, written as a single second-order differential equation, is

$$\{d^2x/dt^2\} + @[1 - x^2]\{dx/dt\} + W^2x = 0 \qquad (6.63)$$

This is known as the *Van der Pol equation*. The Van der Pol equation can be interpreted as a mass-spring system in which the frictional force has the curious property that although energy is dissipated for large displacements from equilibrium, this energy is pumped back into the system when the displacement is small. Thus, it is the interplay between the destabilizing effect of excitation for small displacements and the damping effect which occurs for large displacements that give rise to self-sustained oscillations.

For the study of self-sustained oscillations, Van der Pol's oscillator is regarded to play a special part in biology. It provides an example of a nonlinear physical oscillator. Several models of biological rhythms, such as the contraction of the heart muscle, have been constructed on such a model. This equation has some similarity with the linear homogeneous differential equation

$$\{d^2x/dT^2\} + 2A\{dx/dt\} + w^2x = 0 \qquad (6.64)$$

Equation 6.64 has the solution

$$x(t) = x(o)[\exp(-At)]\sin(pt + q)$$
$$p^2 = w^2 - A^2 \qquad (6.65)$$

When $A < 0$, the amplitude increases, while when $A > 0$, the amplitude decreases and eventually approaches zero. When A equals zero, the case of the harmonic oscillator is obtained.

Equation 6.63 can now be compared to Equation 6.64. For small values of x, that is, when $@[x^2 - 1] < 0$, the amplitude will increase, similar to the situation when A is negative. For large values of x, that is, when $@[x^2 - 1] > 0$, the amplitude will decrease, similar to the positive values of A of Equation 6.63. Therefore, self-sustained oscillations will arise. Corresponding to the situation $A = 0$, the mean value of $@[x^2 - 1]$ will be zero.

Though there are similarities, the curve form of the solutions of Equation 6.63 will be different from the sine curve. For large values of $@$ much larger than unity, the curve will be close to a square form. For larger values of $@$, the time to reach stationary-state oscillations will be shorter. Therefore, the magnitude of $@$ is important in determining the transients.

It can be shown (with the use of the Poincaré-Bendixon theorem), that Equation 6.66a has a periodic orbit, with $@$ being a positive constant. Equation 6.63 written in an equivalent form,

$$\{d^2x/dt^2\} + @[x^2 - 1]\{dx/dt\} + x = 0 \qquad (6.66a)$$

is equivalent to the system of equations

$$\{dx/dt\} = y + @\{x - (x^3/3)\} \qquad (6.66b)$$

$$\{dy/dt\} = -x \qquad (6.66c)$$

This is not the usual phase-plane equivalent of Equation 6.66a. The origin is the only critical point of Equation 6.66 and it is unstable. The trajectories of Equation 6.66 are symmetrical with respect to the origin. The direction field defined by Equation 6.66 may be expressed in the form

$$\{dy/dx\} = -[x/\{y + @[x - (x^3/3)]\}] \qquad (6.67)$$

Thus, in particular, a trajectory becomes horizontal only as it crosses the y-axis and becomes vertical only as it crosses the cubic

$$y = @[(x^3/3) - x] \qquad (6.68)$$

The cubic curve is presented in Figure 6.4. Each nontrivial trajectory, which intersects the cubic curve for values of $x > 3^{(1/2)}$, is presented in Figure 6.4.

The quantity E, proportional to the energy, satisfies the differential equation

$$E = \{x^2 + y^2\}$$
$$\{dE/dt\} = -2@x[(x^3/3) - x] \qquad (6.69)$$

along each trajectory of Equation 6.66. Thus, E decreases along any particular portion of the trajectory lying to the right of the dotted line, $x = (3)^{(1/3)}$.

For small values of $@$, the periodic orbit of the Van der Pol equation is very nearly a circle. This is evident when one examines the direction field of Equation 6.66. As $@$ decreases, Figure 6.4a flattens and for $@$ equal to zero, the cubic passes to the x-axis. For large values of @, the orbits obtained are presented in Figure 6.4b. For large values of @, the corresponding orbit is a very jerky kind of motion. This is called the *relaxation oscillation*.

An oscillator of the Van der Pol kind can be pictured in the phase plane, as presented in Figure 6.4. The trajectories are different from the trajectories of a harmonic oscillator (see Figure 6.2) in one important

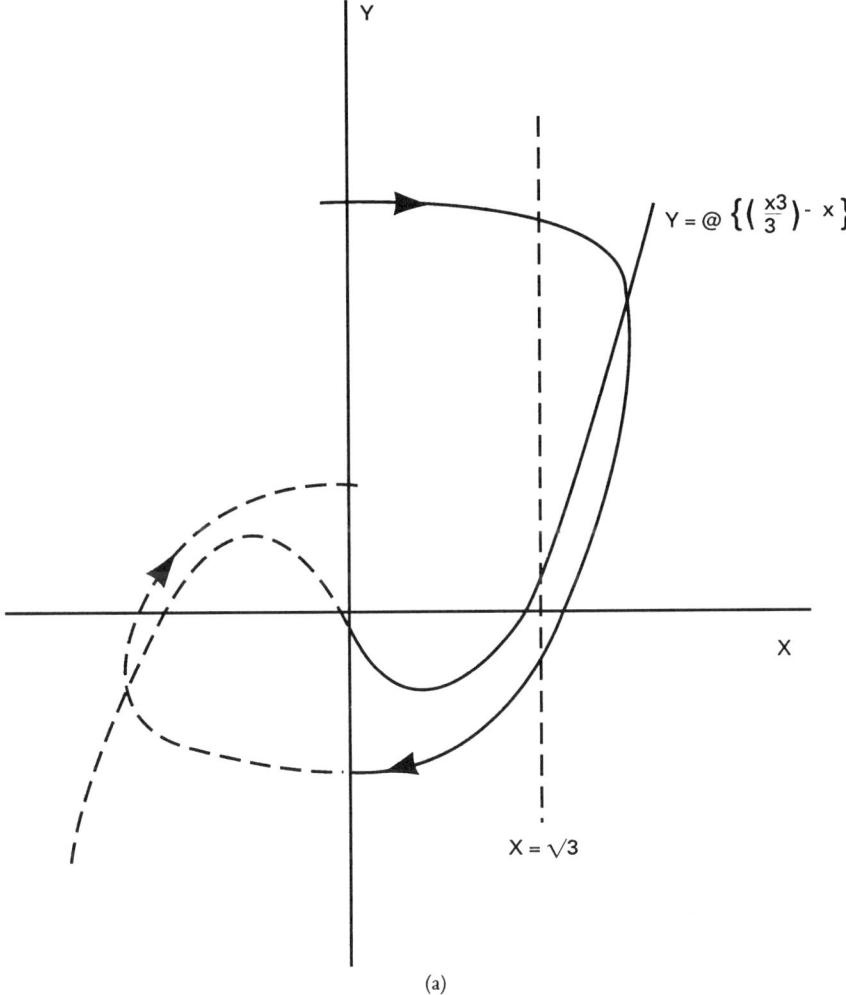

(a)

Figure 6.4 Relaxation oscillation depicted by Van der Pol oscillator. Phase-plane trajectories of Equation 6.69. Note the trajectories are symmetrical with respect to the origin. The quantity, E, proportional to energy, decreases along any portion of trajectory lying to the right of the line $x = 3^{1/2}$. The trajectories obtained for Equation 6.66a for large value of @ are presented in Figure 6.4b. (Reproduced with permission from Struble).

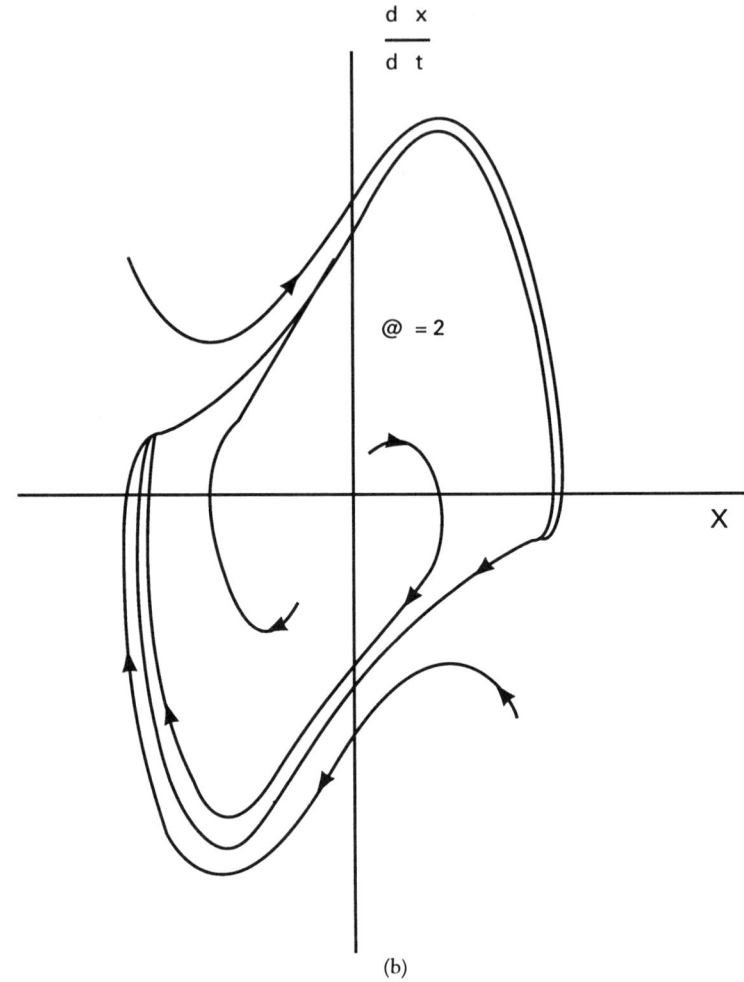

(b)

Figure 6.4 (*continued*)

aspect. When this oscillator is perturbed slightly and moves somewhat away from the closed trajectory, it will always come back to it. The closed trajectory is therefore a stable limit cycle. Linear systems, exemplified by the harmonic oscillator, do not have a limit cycle. The Van der Pol oscillator provides an example of a nonlinear physical oscillator. It has been the basis of construction of several models of biological rhythms.

THE FORCED OSCILLATIONS OF THE VAN DER POL EQUATION

Consider the equation

$$\{d^2x/dt^2\} + x = @[1 - x^2]\{dx/dt\} + @k\cos(ut) \tag{6.70}$$

where each of k and u are positive definite and fixed constants. The constant $@ > 0$ is assumed to be small. The forcing function is considered soft in this case, due to the presence of the small factor $@$. An asymptotic solution of the form in Equation 6.69 can be anticipated.

$$x = A \cos(t - \theta) + @x_1 + @^2 x_2 + \ldots$$

$$\ldots + @^n x_n \tag{6.71}$$

The principal part of Equation 6.71 reflects the basic characteristics of the free oscillations. If the input frequency u is appreciably different from unity, the first-order variational equation can be written as

$$\{d\theta/dt\} = 0$$

$$\{dA/dt\} = (@/8)A[4 - A^2] \tag{6.72}$$

and for the first-order perturbational equation,

$$\{d^2 x_1/dt^2\} + x_1 = (A^3/4) \sin 3(t - \theta) + k \cos ut \tag{6.73}$$

From these one obtains the first-order solution,

$$x = A \cos(t - \theta) - @[A^3/32] \sin 3(t - \theta)$$

$$+ \{(@k)/[1 - u^2]\} \cos ut \tag{6.74}$$

$$A^2 = 4/\{1 + [(4/A_o^2) - 1] \exp(-@t)\}$$

A_o denotes an arbitrary nontrivial value for A. This solution contains both the forced and natural frequencies superimposed as though the system were linear. Since the magnitude of A approaches 2, the natural response can be expected to dominate.

This is not an unexpected result, since Equation 6.72 is *nearly linear* and the forcing function is *soft*. However, as the forcing-function frequency u approaches unity, the character of the solution is modified radically and a somewhat unexpected phenomenon takes place. As u approaches unity, the forced response becomes more significant. But instead of a persistence of both the natural and forced responses, the natural frequency becomes entrained by the forced oscillation.

The result is a synchronization of the output as the input frequency. The soft forcing function commands a response at the input frequency only under a resonant condition with an input frequency nearly equal to the natural frequency. Thus, entrainment phenomenon is perhaps not so unexpected. In general, when similar active elements come into contact with each other to form an extended field, they are expected to produce a wealth of wave patterns and turbulence-like phenomena.

THE CASE OF A SINGLE OSCILLATOR SUBJECT
TO A PERIODIC FORCE _____

If the natural period of a single oscillator is T, and the perturbation $@p$ represents a weak periodic force of period $T^\star$, one has

$$M(f, t) = M(f + T, t) = M(f, t + T^\star) \qquad (6.75)$$

If the difference $[T - T^\star]$ is small and is of the order $@$;

$$1 - [T/T^\star] = @d \qquad (6.76)$$

it is expected that the oscillator comes to oscillate with exactly the same frequency as the external one, if the magnitude of d is below some critical value $d^\star$. More general entrainment in which some integer multiples of T and $T^\star$ become identical could be analyzed in a similar manner.

Let V denote a new phase variable defined by

$$f = [T/T^\star]t + V \qquad (6.77)$$

It is evident that if V is constant in time, this implies that the oscillator is entrained to the external periodicity. The equation for V now becomes

$$\{dV/dt\} = @\{d + M[T/T^\star]t + V, t]\} \qquad (6.78)$$

Equation 6.69 shows that V is a slowly varying function of t. On the other hand, the quantity M, if viewed as a function of t and V, is $T^\star$ periodic in t (and T periodic in V). Since the slow variable V could hardly change during the period $T^\star$, one may safely time average M over this interval with V kept constant. In this manner one obtains, after setting $@ = 1$,

$$\{dV/dt\} = d + G(V)$$

where

$$G(V) = G(V + T)$$

$$= (1/T) \int_o^{T^\star} M\{(T/T^\star)t + V,t\} \, dt \qquad (6.79)$$

Entrainment to the external periodicity occurs if Equation 6.79 has a stable equilibrium solution. A given equilibrium solution V_o is stable if $\{dG/dV\}$ evaluated at $V = V_o$ is negative, and is unstable if it is positive.

When the condition for the entrainment is not satisfied, the oscillator gains a frequency different from the external one. The generalization of the preceding method for a pair of oscillators with different frequency distributions is presented by Kuramoto (1984).

FIELDS OF OSCILLATORS _____

The nonlinear dynamics of dissipative systems are described by the set of first-order ordinary differential equations,

$$\{dX_i/dt\} = F_i(X_1, X_2, \ldots, X_n; u)$$

$$i = 1, 2, \ldots, n \tag{6.80}$$

which include some parameters represented by u. One may write the set of equations in Equation 6.80 in the vector form

$$\{dX/dt\} = F(X; u) \tag{6.81}$$

Let $X^o(t)$ denote a linearly stable T-periodic solution of an n-dimensional system of ordinary differential equations in Equation 6.80. [The set of Equations 6.80 is assumed to admit a stable limit-cycle solution $X^o(wt)$, with period $T = (2\pi/w)$].

$$\{dX^o/dt\} = F(X^o)$$

$$X^o(t + t^\star) = X^o(t) \tag{6.82}$$

If this system is perturbed as

$$\{dX/dt\} = F(X) + @p(X) \tag{6.83}$$

where $@p(x)$ represents a small perturbation, generally depending on X, and $@$ is an indicator or the smallness of p and is finally set equal to unity, periodic motions will persist when the perturbation $@p$ is introduced. Its period, however, will deviate slightly from T. This change in period needs to be expressed to the lowest order in $@$. Let C denote the closed orbit corresponding to $X^o(t)$. Since C is considered stable, each state point X in the vicinity of C approaches C as t tends to infinity in the absence of perturbation.

This asymptotic periodic motion on C is described by association of a certain value f to each X contained in C in such a way that the motion on C may produce a constant increase in f.

$$\{df(X)/dt\} = 1 \tag{6.84}$$

for X contained in C. The quantity f is the phase defined on C. Its value is only determined to an integer multiple of T. The definition of phase is not confined to C. For arbitrarily small perturbations which can kick the state point out of C, without the definition of f outside of C, one could no longer say anything about the phase of perturbed oscillations.

Kuramoto (1984) identifies a phase-dependent sensitivity, which measures how sensitively the oscillator responds to external perturbations. When one averages with respect to rapidly oscillating processes, care should be exercised, since this point turns out to be crucial in the identification of phase-description synchronization. The reader is referred to the monograph of Kuramoto (1984) for further details on this subject. These additional materials are too abstract to be included in this chapter.

When one deals with the dynamics of chemical reaction systems, X of Equation 6.80 represents a set of concentrations of the chemical species involved, and u may be taken to be the flow rate at which certain chemicals are fed into the system so that their consumption due to the reactions are compensated.

For some range of u, the system may be stable in a time-independent state. This is usually true if the system is close to thermal equilibrium. In many systems, such a steady state loses stability for some critical value u_c of u. Beyond this, periodic motion exists. This appears as the branching of time-periodic solutions, from a stationary solution branch, in the parameter-amplitude plane. This phenomenon is called the *Hopf bifurcation*. In chemical reactions, the corresponding phenomenon is the onset of chemical oscillations. As u increases further, the system may show more and more complicated dynamics through a number of additional bifurcations. Complicated periodic oscillations, quasi-periodic oscillations, as well as a variety of nonperiodic (chaotic) behaviors may appear. This appears to be the basis of various dynamical complex behaviors exhibited by systems.

Consider the reaction–diffusion system,

$$\{\partial X / \partial t\} = F(X) + D \, \nabla^2 X \tag{6.85}$$

where D is a matrix of diffusion coefficients. The stability of the uniform steady-state solution X_o of Equation 6.85 can be analyzed from the variational equations about X_o.

$$\{\partial u / \partial t\} = L + D\{\partial^2 / \partial x^2\}u \tag{6.86}$$

This equation admits solutions of the form

$$u(x, t) = V \exp(mt) \cos[h\pi / d]x \tag{6.87}$$

By substitution, one has the eigenvalue problem

$$L_h V_h = m V_h$$
$$\tag{6.88}$$
$$L_h = L - D\{h\pi / d\}^2$$

For vanishing values of h, this reduces to the eigenvalue problem

$$LU = mU \tag{6.89}$$

When X and F are n-dimensional real vectors and u is a real scalar parameter, $X_o(u)$ denotes a steady-state solution of Equation 6.85; that is,

$$F\{X_o(u); u\} = 0 \qquad (6.90)$$

When one expresses the deviation of X from X_o by U and expands in a Taylor series,

$$\{dU/dt\} = LU + MUU + NUUU + \dots \qquad (6.91)$$

the expansion coefficients generally depend on u at least through $X_o(u)$.

The stability of X_o is related to the distribution of the eigenvalues m of Equation 6.90 in the complex plane. Each of these eigenvalues forms a branch, which is continuous if the system length d is very large. In this manner, the reaction–diffusion equation system can be reduced to a system of ordinary diffusion equations of the kind

$$\{dC_{ah}/dt\} = l_{ah}C_{ah} + \text{nonlinear terms} \qquad (6.92)$$

In excitable reaction-diffusion systems, pulses can travel as a periodic wave train. In oscillatory reaction-diffusion systems also, the existence of plane-wave solutions has been established theoretically. When pulses (or peaks of plane waves) are distributed to form an aperiodic train, they are expected to regulate themselves to reestablish a perfect periodicity.

This concludes our presentation of mathematical aspects of periodic phenomena and stability aspects. A few more additional aspects are presented in Chapter 8 on rhythmic phenomena.

REFERENCES

BORN, M., and K. HUANG, *Dynamical Theory of Crystal Lattices.* London: Clarendon Press, 1954.

BELTRAMI, E., *Mathematics for Dynamic Modelling.* New York: Academic Press, 1987.

CHANDRASEKHAR, S., *Plasma Physics,* notes compiled by S.K. Trehan. Chicago: The University of Chicago Press, Phoenix Science Series, 1960.

DAVIS, H.T., *Introduction to Nonlinear Differential and Integral Equations.* New York: Dover Publications, 1962.

KURAMOTO, Y., *Chemical Oscillations, Waves, and Turbulence.* Berlin: Springer-Verlag, 1984.

MARGENAU, H., and G.M. MURPHY, *The Mathematics of Physics and Chemistry,* vol. 1. New York: Van Nostrand, Inc., 1943.

MINORSKY, N., *Nonlinear Oscillations.* New York: Van Nostrand Co., 1962.

MORSE, P.M., and H. FESHBACH, *Methods of Theoretical Physics.* New York: McGraw-Hill Book Company, 1953.

STRUBLE, R., *Nonlinear Differential Equations.* New York: McGraw-Hill Book Company, 1962.

chapter 7

Some Control Mechanism Models

Conventionally, scientific laws have been constructed in terms of a particular set of mathematical functions and constructs. They have often been developed for their mathematical simplicity as for their capacity to model salient features of the system behavior. For a physiological system, the modeling is a concise way of quantifying the interaction behavior of those physiological mechanisms which have a more or less clearly defined functional goal.

Modeling comprises at least two major approaches. The first one is writing equations for well-established physical processes from a priori knowledge of the equations of conservation, continuity, or balance. Although the physical laws may be known, this step is not trivial and simplifying assumptions may be necessary. The second one is an empirical one based on experimental information regarding the behavior of the system. Neural and hormonal control phenomena are often modeled by the empirical approach (Stear and Kadish, 1969). Models are most useful when they fail, for then they may indicate the need to look for missing elements. Psychologically, however, they are most appealing when they succeed in explaining the data. Models with negative results are rarely published.

Several models have been proposed by various investigators to describe different control mechanisms suspected to occur in biological systems. Only a selected few of these are presented in this chapter, in order to save space and to illustrate the general method and philosophy adopted. Mathematical modeling of physical and biological processes is associated with the establishment of reasonable mathematical representations, which relate the inputs to the outputs. A model is an idealized representation of reality. Models of living systems are of necessity simplifications, in which certain features of the system are examined. This simplification is an idealization and an abstraction that allows one to make some sense of what would otherwise be a mass of confusing information. The model should never be confused with the system it represents. In connection with modeling, some of the following questions are usually raised.

To what extent is it possible to obtain insight into the internal structure of the system from input–output measurements?

Can one obtain structural identifiability?

What experiments are necessary to determine the internal connections uniquely? An optimal experiment is one in which the greatest amount of information on the parameters of the system can be determined. In general, it may not be possible to measure each state variable in a particular problem. In this connection the following definitions are useful.

- *Controllability*: Controllability of the system means the possibility of influencing independently each state of the system through the inputs.
- *Observability*: Observability means the possibility of reconstructing each state of the system from the outputs.
- *Reachability*: Reachability implies the possibility of the transfer of the system from some current state, using a given set of inputs, to a new prescribed state within a certain amount of time.
- *Identifiability*: Identifiability implies that the system's unknown parameters can be uniquely determined from experimental results.

Consider the control problem

$$\{dx/dt\} = f(x, y, u)$$
$$\{dy/dt\} = g(x, y, u)$$

with the initial conditions,

$$x(t = 0) = x_o$$

and

$$y(t = 0) = y_o \qquad (7.1)$$

The values of x and y are deemed positive definite. u is called the control variable, which is bounded. The functions f and g are in general nonlinear in x, y, and u. These equations permit the formulation of a single differential equation, connecting x and y.

When $u = 0$,

$$\{dy/dx\} = \{[g(x, y, 0)]/[f(x, y, 0)]\}$$
$$y(x_o) = y_o \tag{7.2}$$

The trajectories of this equation in the phase plane represent the situation when there is no control.

The two major systems involved in control of all important physiological variables and processes are the central nervous system and the endocrine system. In living systems, the metabolic processes cause the formation of molecules, transport, and elimination of materials from the system. Blood circulation, which constitutes the transportation system between locations of production of materials and cells in which such materials are utilized, constitute a well-developed and integrated system of control. Hormones are produced in a particular endocrine gland and are present in the blood only in very small amounts. Models of endocrine-metabolic systems are based usually on physical principles and hypotheses about the structure of the system and possibly its functioning. Available a priori knowledge is utilized in formulating the model, and state variables tend to have a direct counterpart in the system.

We now present several models, the first of which is a model for genetic control.

MODEL 1: A MODEL FOR A GENETIC CONTROL MECHANISM

Feedback repression has been considered to be a general basis for biological regulation. The demonstration of a genetic feedback repression in enzyme induction started considerable interest in the analysis of the models of biological regulation.

Consider a reacting chemical system, open to the environment. Let there be three substances participating in the chemical reactions occurring. Let these be x, y, and z. The reactions occur subject to the following constraints.

1. x increases at a rate inversely proportional to z and is removed from the system at a constant rate to an external sink.
2. y increases at a rate directly proportional to x and is removed at a constant rate to an external sink.
3. z increases at a rate directly proportional to y and is removed at a rate proportional to its own concentration.

These reactions are described by the following rate equations.

$$\{dx/dt\} = [K/z] - p \tag{7.3a}$$

$$\{dy/dt\} = Lx - q \tag{7.3b}$$

$$\{dz/dt\} = My - rz \tag{7.3c}$$

If the concentration of z is very large, such that its variation with time can be neglected, Equations 7.3 simplify to

$$\{dx/dt\} = [K/Jy] - p$$
$$\{dy/dt\} = Lx - q \tag{7.4}$$
$$J = [M/r]$$

A Hamiltonian representation can now be made. Hamilton's equations for this system would be

$$[\partial H/\partial y] = [K/Jy] - p$$
$$-[\partial H/\partial x] = Lx - q \tag{7.5}$$

Upon integration, these equations yield

$$H(x, y) = \{K/J\} \ln y - py - (L/2)x^2 + qx \tag{7.6}$$

The Hamiltonian formulation indicates that one can define an integral for metabolic reactions which will serve the same function in the study of cellular activity that the energy integral plays in classical mechanics.

The set of equations can be written as a single second-order differential equation:

$$\{d^2y/dt^2\} - L\{[K/Jy] - p\} = 0 \tag{7.7}$$

Introduction of new state variables, $g = \{dx/dt\}$, enables one to rewrite Equation 7.7 as a pair of dynamical equations in the state variables, y, g:

$$\{dy/dt\} = g$$
$$\{dy/dt\} = L\{[K/Jy] - p\} \tag{7.8}$$

Exhibited in this manner, the state variable y plays the role of a mechanical displacement, and g is its corresponding momentum. Regarded as a conservative mechanical system, the Hamiltonian for the system is given by

$$H(y, g) = \{g/2\} - \{LK/J\} \ln y + qy \tag{7.9}$$

Now if x represents the concentration of a primary gene product, y represents the concentration of a particular enzymatic protein determined by the gene, and z is the concentration of a metabolite produced as a consequence of the activity of the protein, the set of mathematical equations for the model represents a *genetic control mechanism* (Goodwin, 1973).

Goodwin (1965) modified the set of Equations 7.3 to include a term for a time lag, in the first as

$$\{dx/dt\} = [a/(A + kz)] - px \tag{7.3d}$$

Analog computer simulation of the set of Equations 7.3b, 7.3c, and 7.3d led to the conclusion that this was an oscillatory system. Griffith (1968) concluded that undamped oscillations can never occur for any values of the parameters of these sets of equations, when repression is accomplished by a protein. Such oscillations are possible when repression is due to the metabolite but only when there is a cooperative repression of very high order.

Fraser and Tiwari (1974) have concluded that single gene loops without stochastic components will show undamped oscillations under very restrictive conditions that are unlikely to occur in natural systems. Single gene loops which include stochastic components will show undamped oscillation, but such oscillation will be irregular in both period and amplitude. The life

---

† A dynamical system of n degrees of freedom can be described by the set of generalized coordinates $q_1, \ldots, q_n$. The momentum p_j is associated with or conjugate to position coordinates q_j. The momentum p equals mass times velocity. The Hamiltonian H is related to p_j and q_j by the canonical relations

$$\{\partial H/\partial q_j\} = -\dot{p}_j$$
$$\{\partial H/\partial p_j\} = \dot{q}_j$$

If F and G are two functions of the canonical variables, p_i's and q_i's, the Poisson bracket of F and G $\{F, G\}$ is defined by

$$\{F, G\} = \sum_i \langle\{\partial F/\partial q_i\}\{\partial G/\partial p_i\} - \{\partial F/\partial p_i\}\{\partial G/\partial q_i\}\rangle$$

Thus, the time derivative of an arbitrary function F which is a function of momenta and coordinates can be written as

$$\{dF/dt\} = \{\partial F/\partial t\} + \sum_i [\{\partial F/\partial q_i\}\dot{q}_i + \{\partial F/\partial p_i\}\dot{p}_i]$$

With Hamiltonian formulation, this becomes

$$\{dF/dt\} = \{\partial F/\partial t\} + \{F, H\}$$

This is a very concise way of writing the dynamical equations of motion for the system. It provides a test for recognizing constants of motion. If

$$\{\partial F/\partial t\} = -\{F, H\}$$

F is evidently a constant of motion. If F does not explicitly depend on time, it is a constant of motion if its Poisson bracket with the Hamiltonian vanishes. This can be beneficially exploited for many biological model problems.

Lagrange's equations are a system of $3N$ second-order differential equations for the $3N$ generalized equations. In the Hamiltonian formulation, an additional set of $3N$ independent variables is introduced. This leads to $6N$ first-order differential equations, describing the motion of the system.

A system of n degrees of freedom will have n differential equations, which are second order in time. The solution of each equation will require two integrations resulting in $2n$

span of protein specifies the period of any oscillation, but has little, if any, effect on whether the oscillation is damped or undamped.

MODEL 2: A MODEL FOR INSULIN RELEASE AND ACTION

Insulin is a hormone formed in the pancreas. It decreases liver and muscle glycogenolysis and facilitates glucose transfer into the cells. In this manner, blood glucose is inversely related to insulin level. Glucagon, which is also a pancreatic hormone, increases liver glycogenolysis. When the glucose concentration in blood rises, pancreatic secretion of insulin is stimulated. Depression of muscle and liver glycogenolysis and increased glucose oxidation lead to a lowering of blood glucose. A fall in blood glucose leads to the secretion of glucocogon and an increase in blood glucose. In this manner, insulin and glucagon appear to play a reciprocal role in the regulation of blood glucose level. The principal mechanisms appear to be the increased glycogen synthesis caused by insulin and the increased glycogenolysis caused by glycogon. There are of course other metabolic actions attributable to insulin. At the molecular level, insulin appears to have a direct effect on glycogon synthesis in addition to an effect on membrane permeability to glucose.

---

constants of integration. In a specific problem these will be determined from the initial conditions. It is not always possible to integrate such equations of motion. Even when complete solutions cannot be obtained, it is often possible to extract additional information about the physical nature of the system motion.

Associated with every mechanical system, there is a certain kinetic energy T, and if the system is conservative, then there is a potential energy V. The kinetic energy is a function of positional coordinates and velocities, while the potential energy is in general a function of position coordinates only. Hamilton's principle of least action essentially states that the motion of the system between any two times takes place in such a way that the definite integral of $L = T - V$ between these two times has an extremum value.

These 2 n partial differential equations of the first order are known as the *canonical equations of motion*.

The time rate of change of H is

$$\dot{H} = \sum_{j}^{n} \{\partial H / \partial q_i\}\dot{q}_j + \sum_{j}^{n} \{\partial H / \partial p_j\}\dot{p}_j$$

The substitution of the values of q_j and p_j from the Hamiltonian equation gives the result that H is a constant with time. Thus, for conservative systems, since energy is constant, one deduces the connection between energy and the Hamiltonian formulation.

The Hamiltonian formulation gives us a powerful method of working with the physical principles involved. In the Hamiltonian formulation, momenta are considered as independent variables on an equal footing with the coordinates. If the Hamiltonian function is not an explicit function of the time, it is a constant of motion. In the case of a dynamical system, and a coordinate system, such that the time does not appear in the equations defining the generalized coordinates, the kinetic energy is a homogeneous quadratic function of the velocities. If H is independent of a generalized coordinate, the corresponding canonical momentum is a constant of motion.

A continuous model for the interaction of insulin and sugar in the human body has been proposed by Davies. In this model, the amount of sugar in the blood at any time is assumed to change in three possible manners:

1. The sugar is metabolized by the body due to the presence of insulin.
2. The sugar is released into the blood from the liver if the concentration is below some minimum level, g_m.
3. The sugar increases as food is consumed.

The concentration of insulin in the blood undergoes change in two possible ways:

1. Insulin is released into the blood if the amount of sugar is greater than the certain minimum level, g_m.
2. Insulin undergoes degradation in the body, depending on the amount of insulin present in the blood; that is, the higher the concentration, the faster it breaks down.

The system is thus self-regulating, which is an example of a negative feedback loop. If the amount of sugar is greater than g_m, this causes the release of insulin into the blood, which in turn causes the metabolism or uptake of sugar by the cells. The insulin in the blood also degrades, and the system returns to the normal homeostatic condition.

The mathematical analysis of the previously mentioned descriptions can be stated as follows. One has the set of differential equations:

$$\{dG/dt\} = -a_1 GY + a_2[g_m - G]H(g)[g_m - G] + a_3 f(t)$$

$$\{dY/dt\} = b_1[G - g_m]H(g)[G - g_m] - b_2 Y \qquad (7.10)$$

where $G(t)$ is amount of sugar in blood at time t, $Y(t)$ is the amount of insulin in blood at time t, and $H(g)$ is a step function:

$$H(g) = \begin{cases} 1, & \text{when } G - g_m > 0 \\ 0, & \text{when } G - g_m < 0 \end{cases} \qquad (7.11)$$

The last term, $a_3 f(t)$, takes into account the consumption of food, and the resulting increase in sugar concentration. A plausible expression for $f(t)$ is

$$f(t) = \begin{cases} 0, & \text{for } t < t_0 \\ Q \exp\{-k(t - t_o)\} & \text{for } t > t_o \end{cases}$$

where t_o is the time at which food is taken. Q is the value of $f(t)$ at time $t = t_o$. a_i and b_i, $i = 1, 2, 3, \ldots$ are rate constants. $H(g)$ implies that the function H is a function of glucose concentration, which can assume only null and unity values.

The step functions serve numerically to turn on or turn off various terms as g is greater than or less than g_m. The key terms are a_1 and b_1, which couple the two differential equations. In this context, diabetes is characterized by a_1 and/or by b_1 being very low. The metabolism of sugar is assumed to be unaffected by insulin and/or the secretion of insulin into the blood is unaffected by sugar concentraion.

When a_1 and b_1 are very small, and when $G > g_m$, and Y is small, in the equation for $\{dG/dt\}$, only the food consumption term is significant. There is very little stimulus for G to return to the value g_m, since both a_1 and Y are small. The amount of insulin present, $Y(t)$, increases very slowly, and since b_1 is small, this results in very little influence in the reduction of G.

On the other hand, if $G < g_m$ initially, and Y is small, then Y has no source term. But since sugar will be released into the bloodstream, G has now a source term. Regardless of the presence or absence of a food source, G has a method of reaching g_m. Therefore, the problem of the regulation of sugar concentration arises only when G is greater than g_m. Hence, the solution is to add another source term to the differential equation for insulin concentration, say, $b_3 g(t)$. This results in the secretion of insulin whenever sugar is metabolized, thus compensating for low values of a_1.

The secretion of insulin by the perfused pancreas in response to a constant concentrated glucose is multiphasic, with a spike followed by a slow phase, which renders the pancreas hypersensitive to further stimulation. A negative feedback hypothesis and a storage limited two–compartment model with a large storage and a small labile compartment have been proposed to account for the observed phenomena (Guyton *et al.*, 1978). Since the relative constancy of the duration of the spike phase is not compatible with the simple two–compartment model, a threshold distribution model was proposed in which pockets of insulin are related at different glucose concentrations, the numbers of each kind of packet being replenished according to a specific algorithm (Grodsky, 1972).

The model of Landhal and Grodsky (1982) assumes that glucose is a metabolite, and in the presence of calcium ions forms a moiety, X, whose concentration depends on the concentration of glucose. Therefore, $X = X(g)$. It is assumed that $X(g)$ has these two effects: stimulation of the production of a provisionary quantity P, and a dual secretory role. The secretion term is thus made quite complex.

The total quantity of insulin Q available for release at any one time will be determined by the rate of exchange between the storage compartments, the rate of loss due to secretion S, and an incremental rate due to provision. If C is the concentration and V is the total volume, the amount available for release $Q = CV$ is determined by the differential equation:

$$V\{dC/dt\} = K^+C_s - K^-C_s + hPV - S \qquad (7.12)$$

where K^+ and K^- are rate constants and C_s denotes the concentration in the storage compartment (if volume is V_s), assumed constant, for the duration of the experiments. If H denotes $\{K^-/V\}$ and $Q_o = \{[K^+VC_s]/K^-\}$, the stationary state amount G equals zero, the preceding equation becomes

$$\{dQ/dt\} = H[Q_o - Q] + h^\star P - S$$
$$h^\star = hV \qquad (7.13)$$

The quantity P is assumed to be produced at a rate depending on the concentration of glucose, $mP^\star(g)$, and lost at rate mP. Thus, one has

$$\{dP/dt\} = m[P^\star(g) - P]$$
$$P^\star = P(\infty) \qquad (7.14)$$

It is assumed that $P^\star$ is dimensionless so that $h^\star$ will have the units (mg/minute). The simple compartment relation is utilized for the first secretory process. For the second secretory process, which is signal limited, it is assumed to be proportional either to the difference between the outside concentration X and inside concentration X_i, or to the difference between an excitation taken equal to X and an inhibitory entity I, produced at a rate proportional to X, BNX and lost with a rate constant B with N being the proportionality constant. In the later case one may write,

$$\{dI/dt\} = B[NX - I] \qquad (7.15)$$

In the signal-limited concentration difference model, if X is destroyed according to first-order rate kinetics, with a rate constant of m, Equation 7.15 is obtained for I. I is the internal concentration, and N equals the ratio $\{H^\star/[H^\star + p]\}$. $H^\star$ is the rate constant for the exchange across the membrane barrier. N is less than unity for values of p greater than zero. This component of secretion will be q times $\{M_2(X - 1)\}$; that is, the secretion is zero, if the difference is negative. Hence, the total secretion S per minute due to the two processes is given by

$$S = \{M_1Y(g) + M_2(X - I)\}Q \qquad (7.16)$$

Thus, in the model, p acts only to increase the amount of insulin in the labile compartment.

Substitution of the value of S in Equation 7.13 gives a nonlinear differential equation which together with Equations 7.14 and 7.15 can be solved numerically for Q, I, and P as a function of time for a given pattern of glucose stimulation. Thus, the pattern of secretion can be calculated. For a step increase from 0 to g at time $t = 0$, the second term is dominant, so that the early secretion rate tends to be proportional to $X(g)$. The last secretion rate can be obtained from Equations 7.13–7.16 in the steady state and is given by

$$S(g, \infty) = \{h \star P(\infty) + HQ_o\}\{M_1 Y + M_2 RX\}/Z$$
$$Z = [M_1 Y + M_2 RX + H] \tag{7.17}$$

where $R = \{1 - N\}$.

If R and H are small enough, it can be seen that the late secretion rate $S(g, \infty)$ is proportional to $P\star(g)$. By invoking the specific assumption that the two entities of X must form an attachment to some receptor system in order that either P can be produced, or that the second component of secretion can occur, the same earlier results of Grodsky (1972) can be obtained.

Further, if one assumes that four molecules of glucose can combine on some moiety in the presence of calcium ions, and if $X(g)$ is the fraction of having four glucose molecules attached, then X will be given by $\{g \exp (4)\}$ divided by a fourth-degree polynomial in g. If there is strong enough cooperativity, one may be able to neglect all but the zeroth and fourth-degree terms in the denominator and obtain the approximate expression:

$$X(g) = [g^4/\{g_o^4 + g^4\}] \tag{7.18}$$

where g_o is the value of g for which X equals $(1/2)$. This expression with g_o assumed equal to 150 yields values close to those utilized previously.

MODEL 3: THEORIES OF GROWTH

The problem of animal and human growth has attracted many theories. At present we have no theory which would give a satisfactory explanation of the causes and driving forces of animal growth. It was suggested that animal growth is an autocatalytic process and that the velocity of transformation is, at any instant, proportional to the amount of material which is undergoing change and to the amount of material which has been already transformed (Zotin, 1972). Thus, an equation similar to the monomolecular chemical reaction subjected to autocatalysis has been suggested.

$$\{dP/dt\} = k_g P[P_m - P] \tag{7.19}$$

where P denotes the weight of the animal, and P_m is the maximal weight of the animal. t denotes time variable and k_g is a constant called the *growth constant*.

The integral of Equation 7.19 yields the result

$$\ln\{P/[P_m - P]\} = k_g P_m t + C = k_g P_m[t - t_1] \tag{7.20}$$

where t_1 denotes the time necessary to attain half the maximal weight. Equation 7.20 yields an S-shaped curve with an inflection point at the middle and satisfactorily describes the experimental results.

In place of Equation 7.19, if the kinetics of growth is given by an equation analogous to that used for monomolecular chemical reaction, one has

$$\{dP/dt\} = k_g[P_m - P] \qquad (7.21)$$

The solution of Equation 7.21 has the form

$$P = P_m\{1 - e^{-k}g^t\} \qquad (7.22)$$

The general type of the curve describing the weight increase of the animal is assumed to be determined by a single main chemical reaction. The period of growth can be divided into a number of phases, each of which has its own main chemical reaction with constants different from those characterizing other phases. Thus, for each of the growth phases, the following equation is assumed valid.

$$(1/P)\{dP/dt\} = R \qquad (7.23)$$

where R is a constant. The dependence of weight on time is that exponential for each growth phase. The S-shaped curve obtained from experiments is accounted for by the action of forces decelerating the process of weight increase, which begins at a certain instant.

It was suggested that the Gompertz function should be used for the description of the growth. In the differential form, this function looks like

$$(1/y)\{dy/dt\} = ke^{-ax} \qquad (7.24)$$

MODEL 4: THE PROBLEM OF CONFLICT AND PURSUIT ⎯⎯⎯⎯⎯

It is instructive to consider two problems which exhibit a number of difficulties encountered in connection with nonlinear problems. The first problem is the competition between species, as the growth and recession of the populations of two species, one of which preys on the other. This is the *predator-prey problem*. The second problem is concerned with the *curves of pursuit*, in which a path generated by a point P moves in such a manner that its direction of motion is always toward a second point Q, constrained to move along a prescribed path.

The growth of human populations and other organisms presents a characteristic pattern suggesting that it might be described by a single equation. When a population is able to develop freely without obstacles, it grows according to a geometric progression: if the development takes place in the midst of obstacles of all kinds which tend to arrest the growth, and if the social state does not change, the population does not increase indefinitely, but tends to become more and more stationary. This behavior is described by what is known as the *logistic curve*, which is S-shaped (Figure 7.1).

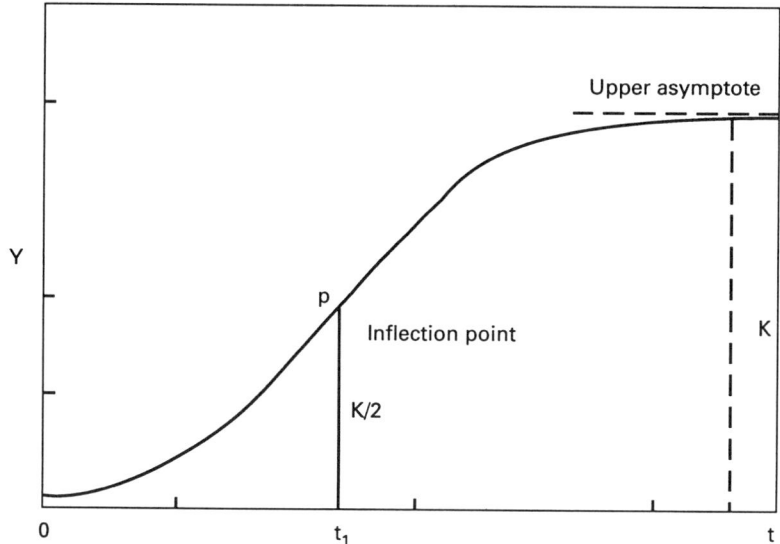

Figure 7.1 The logistic curve.

Assume that the population has an initial size $y(o)$ at time t_o and that after an elapsed time interval and at time t, it has increased to a value $y(t)$. The simplest assumption for the uninhibited growth that can be utilized is that the rate of increase of the population is proportional to the size of the population at that time. Thus,

$$\{dy/dt\} = Ay(t) \tag{7.25}$$

This simple differential equation yields

$$y(t) = y(o) \exp[At] \tag{7.26}$$

as the exponential law of growth. This result may hold evidently only for the initial period of population growth, but cannot exhibit the saturation stationary-state behavior observed over a long period of time.

One may assume that there exists an inhibiting factor for the growth of the population proportional to the square of the population. One may then write a more realistic equation:

$$\{dy/dt\} = Ay(t) - By(t)^2 \tag{7.27}$$

which may be rewritten in the form

$$\{dy/dt\} = My(t)[1 - (y/K)] \tag{7.28}$$

which is a simple null form of a well-known equation called the *Ricati equation*. The solution of Equation 7.28 is

$$y(t) = \{K/[1 + C \exp(-Mt)]\} \tag{7.29}$$

where C is the constant of integration. For positive values of C, an S-shaped curve is obtained from this equation, which is presented in Figure 7.2. This is called the *Logistic Curve*. The curve possesses upper and lower asymptotes. The upper asymptote is the line $y = K$. The curve has one inflection point p, at $t = t_1$ and $y = y_1$.

$$Mt_1 = \log C$$
$$y_1 = K/2 \tag{7.30}$$

A generalization of Equation 7.28 is

$$\{dy/dt\} = P(t)y(t)[1 - (y/K)] \tag{7.31}$$

The solution of Equation 7.31 can be obtained as

$$y(t) = \{K/[1 + C \exp\{g(t)\}]\} \tag{7.32a}$$

where

$$g(t) = K \int^t P(t) \, dt \tag{7.32b}$$

If $g(t)$ is a function which varies continuously between $-\infty$ and $+\infty$ as t varies along a segment of the real axis, between $t = a$ and $t = b$, then horizontal asymptotes exist, which are the lines $y = 0$, and $y = K$. The maxima and the minima of the curve are obtained for values of t which satisfy the condition

$$\{dP/dt\} = 0 \tag{7.33}$$

The second derivative of y of Equation 7.31 is

$$\{d^2y/dt^2\} = [\{dP/dt\} + P^2(t)\{2y - K\}]y(y - K) \tag{7.34}$$

Equation 7.32 belongs to a class of curves defined by the differential equation

$$\{dy/dt\} = G(t)F(y/K)y(t) \tag{7.35}$$

where $F(z)$ is a function such that $F(1) = 0$. The logistic curve and its generalization presented are obtained by setting $F(z) = [z - 1]$.

Another specialization of Equation 7.35 is obtained from the choice

$$F(z) = \log z$$
$$G(t) = \log b \tag{7.36}$$

The solution then is called the *Gompertz curve*:

$$y = KC \exp(bt) \tag{7.37}$$

where C is an arbitrary constant and b is assumed to be less than unity. A suggestion that the description of growth in mammals can be described by an equation of the kind

$$\ln \ln(k/y) = a(b - x) \tag{7.38}$$

where a and b are constants has been suggested. This is actually a solution of the Equation 7.24. The conclusion that the Gompertz function can be utilized for a description of various types of growth is based actually upon experimental results.

The suggestion that the animal growth curve can be divided into a number of regions, for each of which an exponential relation of Equation 7.23 is satisfied, the values of R are different, and can in fact be a function of time. Then the growth can be described by the system of two equations:

$$(1/P)\{dP/dt\} = R$$
$$(1/R)\{dR/dt\} = -a \qquad (7.39)$$

The solution of Equation 7.39 can be written as

$$\ln R = -at + b \qquad (7.40)$$

which when substituted in Equation 7.39 yields

$$(1/P)\{dP/dt\} = e^{(b-at)} \qquad (7.41)$$

Stating that at time $t = 0$, $P = P_o$, the system of Equations 7.39 has the solution

$$P = P_o \exp\{(R/a)[1 - \exp(-at)]\} \qquad (7.42)$$

In these expressions, the relations between flows and fluxes will be nonlinear. The thermodynamic force responsible for growth is not obvious.

MODEL 5: PROBLEM OF GROWTH IN TWO POPULATIONS CONFLICTING WITH EACH OTHER

This problem is a simplified version of a more complicated general problem of society. In many applied physical systems one is concerned with the mutual behavior of two variables x and y, both being functions of time t, which are connected by a system of differential equations of the kind

$$\{dx/dt\} = P(x, y)$$
$$\{dy/dt\} = Q(x, y) \qquad (7.43)$$

It may so happen that cyclical variations in x cause cyclical variations in y, and if the changes in y lag behind those of x, then it is customary to state that there is *hysteresis* in the relationship between them. The problem is illustrated in a simple manner by the following system.

$$\{dx/dt\} = mx - ny$$
$$\{dy/dt\} = px - my \qquad (7.44)$$

where all the parameters are positive definite quantities. These equations state that the growth of both variables is stimulated directly by the magnitude of one of them but is adversely affected by the magnitude of the second. Although this system is linear and its solutions are readily obtained, it serves to illustrate more complicated problems.

In order to find the relationship between x and y, the following equation between them may be studied.

$$px\{dx/dt\} - m[x\{dy/dt\} + y\{dx/dt\}] + ny\{dy/dt\} = 0 \qquad (7.45)$$

Integration of Equation 7.45 yields

$$px^2 - 2mxy + ny^2 = K \qquad (7.46)$$

where K is an arbitrary constant. The quantity $D = nq - m^2$ is defined positive definite, and Equation 7.46 represents an *ellipse*. It is also the solution of the differential equation,

$$\{dy/dx\} = \{[px - my]/[mx - ny]\} \qquad (7.47)$$

Differentiation of Equation 7.44 and elimination of $\{dx/dt\}$ and $\{dy/dt\}$ yields

$$\{d^2x/dt^2\} + Dx = 0 \qquad (7.48)$$

This defines the harmonic

$$x = A\cos\{(Dt)^{(1/2)} + q\} \qquad (7.49)$$

where A and q are constants. In a similar manner, one obtains

$$y = B\cos\{(Dt)^{(1/2)} + r\} \qquad (7.50)$$

With elimination of the variable t, one obtains

$$B^2x^2 - 2AB\cos(q - r)xy + A^2y^2 = A^2B^2\sin^2(q - r) \qquad (7.51)$$

Equation 7.51 is equivalent to Equation 7.46 and defines an ellipse. If q equals r, the ellipse degenerates into two coincidental lines. In Figure 7.2, the two curves obtained from Equations 7.50 and 7.51 are presented for chosen values of the parameters. The ellipse will be called the phase trajectory of the motion and this itself will be described as a vortex cycle about the origin.

The problem of growth of two conflicting populations is a generalization of the problem just described, known usually as the Volterra problem. Let N_1 and N_2 represent the populations of two kinds of species which are in conflict with each other. Assume that the population of the kind 1, denoted as A, preys on the population of the second kind 2, denoted as B. If N_1 is large, then B will flourish in the presence of so much prey, A, and naturally N_2 will decrease. But as N_2 increases, the population of the prey N_1 decreases and starvation will soon set in. Thus, N_2 will start to decrease, and the prey will again begin to increase, since predators are now

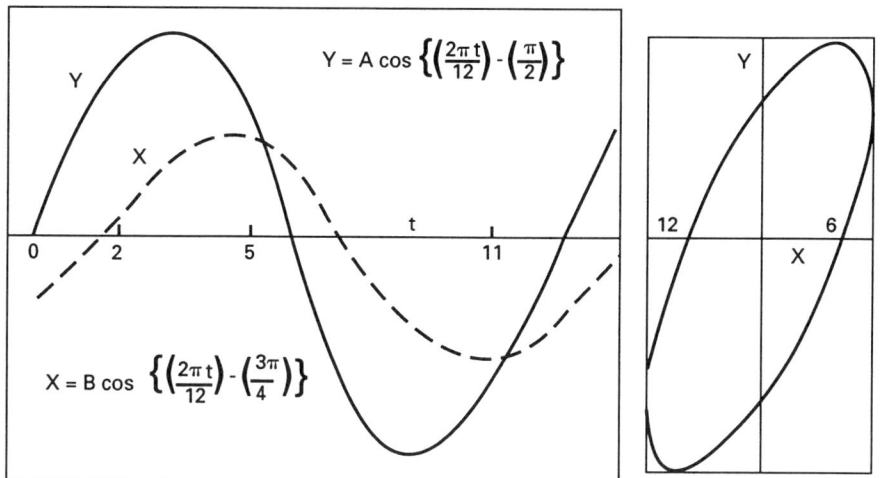

Figure 7.2 Predator-prey relations: The plots of Equations 7.38 and 7.39 plotted for specific values of the parameters. Figure 7.2b indicates the direction of motion of the point through a complete cycle. The ellipse is called the *phase trajectory*.

scarce, and the cycle continues. These kinds of regulation and control of population growth can be formulated in terms of the differential equations

$$\{dN_1/dt\} = aN_1 - bN_1N_2$$
$$\{dN_2/dt\} = -CN_2 + dN_1N_2$$

(7.52)

where a, b, c, and d are positive definite quantities. The number of encounters between members of the two species in a closed space is proportional to the product of their populations. This is the basis of the derivation of Equations 7.52. One should observe that by definition N_1 and N_2 are positive definite. The system of Equations 7.52 has critical points when

$$aN_1 - bN_1N_2 = 0$$
$$-cN_2 + dN_1N_2 = 0$$

(7.53)

Thus, either N_1 or N_2 equals zero, or N_1 equals $\{c/d\}$ and N_2 equals $\{a/b\}$. Linearization about N_1 and N_2 equal to zero leads to the conclusion that the origin is a saddle point. Linerization near $\{c/d\}$, $\{a/b\}$ leads to the conclusion that this critical point is a center. When $N_1 = \{c/d\}$, it follows that $\{dN_1/dt\}$ is positive and N_1 is increasing when $N_2 < \{a/b\}$.

The steady-state conditions yield

$$N_2(o) = [a/b]$$
$$N_1(o) = [c/d]$$

(7.54)

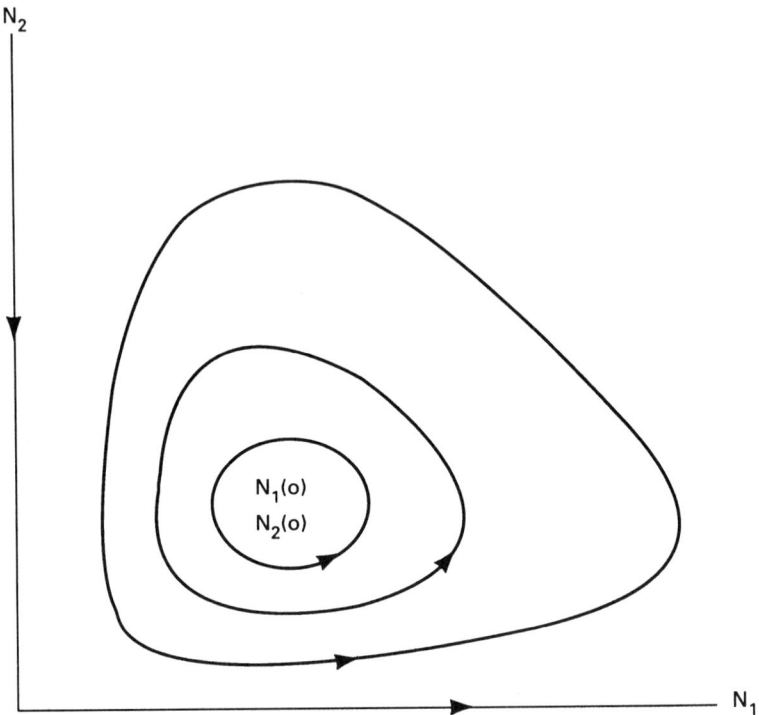

Figure 7.3   Phase-plane plot for a predator-prey problem.

The phase-plane trajectory for the predator-prey problem is presented in Figure 7.3. A closed trajectory represents an oscillation, and unless the values of N_1 and N_2 are zero at time $t = 0$, a maximum (or a minimum) occurs in N_2 about a quarter of a period after the maximum (or minimum) occurs in N_1. An interpretation of this is that the populations of the prey and predator oscillate about their mean values. A change in the environment may switch the system from one trajectory to another. However, periodic fluctuations will continue with no tendency toward the equilibrium state exhibited.

Thus, one may express

$$N_1 = N_1(o) + xe^{iwt}$$
$$N_2 = N_2(o) + ye^{iwt}$$
$$w = \{ac\}^{(1/2)}$$
$$[\x\/\y\] = \{b/d\}\{c/a\}^{(1/2)}$$

$$(7.55)$$

If N_1 represents the population of rabbits and N_2 represents the population of foxes, it is evident that the populations of rabbits and foxes will undergo periodic oscillations of equal frequency, shifted only in phase. In this case, the frequency is amplitude dependent. It is also evident that there are an infinite number of periodic orbits around the steady state.

If one expresses $N_1 = \{cx/d\}$; $N_2 = \{ay/b\}$, Equation 7.53 may be expressed as

$$\{dx/dt\} = a[x - xy]$$
$$\{dy/dt\} = -c[y - xy] \tag{7.56}$$

Differentiation of both expressions and elimination of y and $\{dy/dt\}$ yields the following nonlinear differential equation for $x(t)$.

$$x\{d^2x/dt^2\} = \{dx/dt\}^2 + acx^2$$
$$- cx\{dx/dt\} + cx^2\{dx/dt\} - acx^3 \tag{7.57}$$

Similarly, the elimination of x and $\{dx/dt\}$ yields the nonlinear differential equation for $y(t)$:

$$y\{d^2y/dt^2\} = \{du/dt\}^2 + acy^2 + ay\{dy/dt\}$$
$$- ay^2\{dy/dt\} - acy^3 \tag{7.58}$$

Neither Equation 7.57 nor Equation 7.58 can be integrated in terms of any elementary functions. However, the equations for phase trajectories can *however* be obtained, which is the solution of the equation,

$$\{dy/dx\} = -\{c[y - xy]\}/\{a[x - xy]\} \tag{7.59}$$

Since x and y satisfy Equations 7.52, one can write

$$c\{dx/dt\} + a\{dy/dt\} - (c/x)\{dx/dt\} - (a/y)\{dy/dt\} = 0$$

Integration of this yields,

$$cx + ay - c \log x - a \log y = K \tag{7.60}$$

where K is an arbitrary constant.

Equation (7.60) can be written in a more useful manner.

$$x^{-c}e^{cx} = Cy^ae^{-ay}$$
$$C = e^K \tag{7.61}$$

A functional relationship between x and y is therefore available.

In place of the exponential growth of prey assumed in the absence of the predator, one may assume that the growth is logistic, so that one has the set of differential equations

$$\{dN_1/dt\} = AN_1 - FN_1^2 - BN_1N_2$$
$$\{dN_2/dt\} = -CN_2 + DN_1N_2 \tag{7.62}$$

Evidently, when F equals zero, these reduce to Equations (7.62). One may propose variations of the Lotka–Volterra equations to include other facets of species interactions.

A generalization of the competition between species problem can be written as

$$\{dx/dt\} = F + Cx + Dy + Gx^2 + Hxy + Ky^2$$
$$\{dy/dt\} = E + Ax + By + Lx^2 + Mxy + Ny^2 \qquad (7.63)$$

The trajectories of these equations are closed paths and, therefore, the functions $x(t)$ and $y(t)$ are periodic. The motion can thus be described as stable. If there are more than one kind of prey, and if there are more than one kind of predator, such situations can also be described by further generalizations of these kinds of equations.

MODEL 6. ANTIBODY-TUMOR INTERACTIONS _____

Nonlinear systems are capable of exhibiting time-ordered behavior. There is a family of solutions. Any perturbation away from the stationary-state solution leads to sustained oscillations. The system is structurally unstable. Any small perturbation changes the phase portrait at least qualitatively.

A transition process in an open system implies the establishment of the stationary state following a perturbation. The transition process may be initiated by the action of external or casual internal factors. After the transition, the system can return to the stationary state without undergoing qualitative changes, or it may bring a new stationary state or even to a new steady orbital. There is a similarity between these transitions in open systems and the phase transitions of equilibrium systems. Such transitions can be divided into two kinds. In the first kind, the extensive parameters of the system (concentrations of species, entropy production, and so on), are changed abruptly. In the second kind, only the symmetrical properties of the system are changed abruptly.

The immune system in an organism has many stationary states (orbitals). Antigens influence the transition processes between stationary states. The fight of an organisms's immune system against, say, cancer cells can be described by the following model.

The number of malignant cells X and the number of lymphocytes Y vary. There first appears a significant number of lymphocytes, which with tumor growth reduce the immune system effectiveness.

In the model of Vasiliev and Romanovsky 1983, it is assumed that (1) tumor cells are reproduced exponentially; (2) interaction between lymphocytes and cancer cells results in death of cancer cells; (3) the specific rate of reproduction of cancer cells, Y, is given by

$$m = aX - bX^2 \qquad (7.64)$$

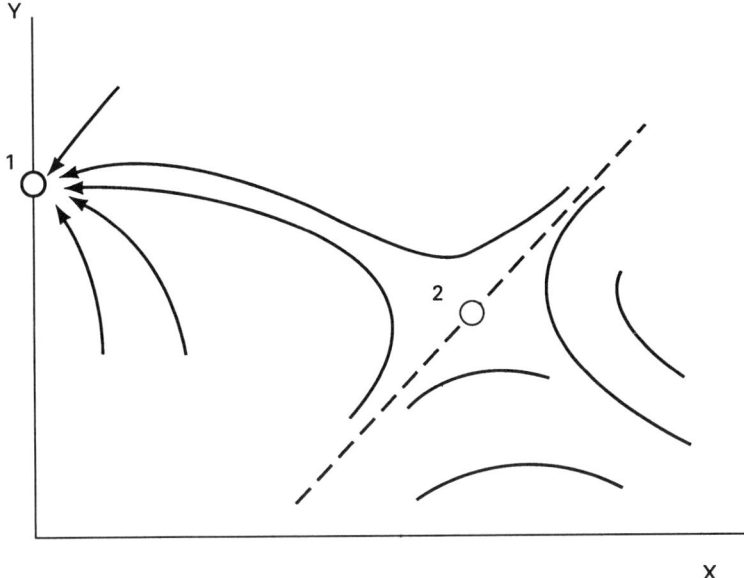

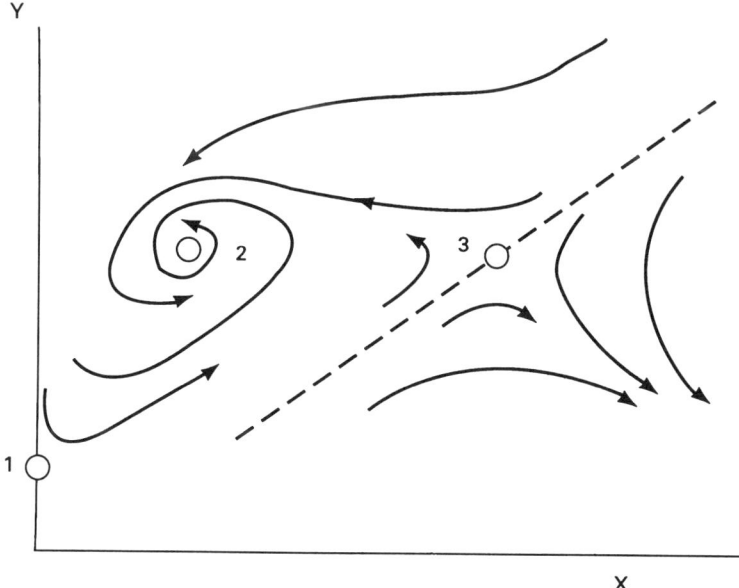

Figure 7.4 Phase-plane pattern for the system of Equations 7.65: Plots of Figure 7.4a are for aperiodic transition processes, and plots of Figure 7.4b are for oscillative transition processes. (Reproduced with permission from Lamprecht and Zotin.)

meaning that at low values of X, the lymphocyte production is stimulated by the tumor, and at high values of X, it is inhibited; and (4) there exists a small stable inflow of lymphocytes with the rate cY. These considerations lead to the set of differential equations describing the model,

$$\{dX/dt\} = aX - bXY$$
$$\{dY/dt\} = aXY - bX^2Y - cY + f - bXY$$

(7.65)

If X is greater than some constant, the system of Equations 7.65 describes two final stationary states, whose phase–plane behavior is presented in Figure 7.4a. The integral curves either approach a fixed point 1, or tend to infinity: X tends to infinity as Y tends to infinity. All the integral curves above the separatrix for the first case are stable orbitals, resulting in recovery (X tending to zero). All the trajectories below the separatrix are stable orbitals of "death." Such interlocation of orbitals explains why, occasionally, stimulation of the immune system with some stimulator inhibits tumor development. But this can occur only in the early stages of cancer, when the value of X is small. Such a curve provides a transition process, bringing to the orbital which leads faster to the recovery or to the transition process through the separatrix which changes the situation.

With some combination of the parameters, another version of the model presents the pattern presented in Figure 7.4b. There are three special points: point 1 in the ordinate is unstable, the integral curves are screwed at the stable point 2. There is no complete recovery in this case, at the end of the fluctuating transition process, an equilibrium between the immune forces and the tumor is established. The system also has two qualitatively different orbitals, different from the previous situation, which leads to either death or partial recovery. Thus, even this simple model enables one to account for the initiation of transition processes resulting in different outcomes in the struggle against tumors. The problem of stable orbitals in growing organisms is analyzed with the properties of dissipative structures (Turing, 1952; Zotin, 1972).

There are many more models for various problems postulated by investigators which are available in the scientific literature. Unfortunately, space does not permit us to present even a significant fraction of these. The five models presented in this chapter are examples of theoretical attempts made in connection with some of the biophysical problems.

REFERENCES

ANDREWS, J.G., and R.R. McCLONE, *Mathematical Modeling.*
BELTRAMI, E., *Mathematics for Dynamic Modeling.* New York: Academic Press, 1987.
CERASI, E., *Quart. Rev. Biophysics,* vol. 8 (1975), p. 1.

DAVIS, H.T., *Introduction to Nonlinear Differential and Integral Equations*. New York: Dover Publishers, 1962.

FRASER, A., and J. TIWARI, *J. Theor. Biol.*, vol. 47, (1974), p. 397.

GOODWIN, B.C., in *Advances in Enzyme Regulation* ed. G. Weber, vol. 3, p. 425, Oxford: Pergamon Press, 1965.

GOODWIN, B.C., *Temporal Organization in Cells*, London: Academic Press, 1973.

GRIFFITH, J.J., *Theor. Biol.*, vol. 20, (1968), p. 202.

GRODSKY, G.M., *J. Clin. Invest.*, vol. 51, (1972), p. 2047.

GRODSKY, G.M., D. CURRY, H. LANDHAL and L. BENNET, *Acta Diabetol. Lat.*, vol. 6, (1969), p. 554.

GUYTON, J.R., R.O. FOSTER, J.S. SOELDNER, M.H. TAN, C.B. KAHN, L. KONCZ, and R.E. GLEASON, *Diabetes*, vol. 27, (1978), p. 1027.

LANDHAL, H., and G.M. GRODSKY, *Bull. Math. Biol.*, vol. 44, (1982), p. 399.

RUBINOW, S.I., *Introduction to Mathematical Biology*. New York: John Wiley & Sons, 1975.

SEGEL, L.A., *Mathematical Models in Molecular and Cellular Biology*. London: Cambridge University Press, 1980.

STEAR, E.B., and A.H. KADISH, *Hormonal Control Systems. Proc. of a Symposium*. New York: American Elsevier Publishing Company, 1969.

TURING, A.M., *Phil. Roy. Soc., London, B*, vol. 237, (1952), p. 37.

VASILIEV, V.A., and YU M. ROMANOVSKY, in *Thermodynamic and Kinetics of Biological Processes*, p. 473, eds. Lamprecht and Zotin. Berlin: Walter de Gruyter & Co., 1983.

ZOTIN, A.I., *Thermodynamic Aspects of Developmental Biology*. New York: S. Karger Publishing Company, 1972.

chapter 8

Nonlinear Nonequilibrium Thermodynamics

PRELIMINARY REMARKS

From a physical point of view, an organism is in the steady state if its individual processes and the entire system maintain a behavior that does not change with time. It was recognized early that biological systems exhibit a relative stability for some modes of behavior. One manifestation of such behavior is a coherent excitation, which implies that a single mode is excited strongly. The stabilization of such a coherent behavior requires nonlinear interactions between various biological subunits. It is intimately connected with cooperative processes. Thermodynamic metastable states and non-equilibrium transitions in such cooperative systems can occur, provided certain energy input is present. Life is characterized by the continuing flow of energy and material. A stationary state may be deemed to have been attained if all possible disturbing factors remain constant in time. Such a steady state is probably artificial and may not characterize the biological system, as it responds to changes in the environment.

The existence of multiple steady states is relevant to regulation and control in biophysical systems. In the thermodynamic sense, an equilibrium state exists only when there is no flow of material or energy. It is equivalent to the terminal state of death. In recent years, a great deal of progress has been made in the study of stability and chaotic behavior of nonlinear systems. Pseudorandom phenomena have been shown to be relevant to a variety of engineering applications in communication and control theory.

In an attempt to bring processes occurring in living systems within the realms of physics and thermodynamics, considerable effort has gone into the development of nonequilibrium nonlinear thermodynamics. The question naturally arises as to whether one can extend the methods of thermodynamics to the range of phenomena where nonlinear situations occur and instabilities exist. Prigogine and Glansdorf (1974) contend that this is possible, so long as the local entropy can be expressed in terms of the same independent variables as if the system were at equilibrium. This is known as the *assumption of local equilibrium*. The entropy production and its second differential are the two basic quantities considered in the formulation of nonequilibrium thermodynamics. A synopsis of essential elements of salient aspects of the nonequilibrium thermodynamics and conclusions of Prigogine and his school (1971) are presented in this chapter. The subject matter covered in this chapter is extremely complicated requiring prior knowledge from many areas of physics, as one may recognize by inspection of the book of Glansdorf and Prigogine (1971) and of Nicolis and Prigogine (1977). Therefore, our coverage of this complicated subject can be only brief and superficial.

It should be stressed that the discipline of thermodynamics provides a reduced description of a macroscopic system. The reciprocal relations of Onsager and the theorem of minimum entropy production belong to the field of linear nonequilibrium thermodynamics. In general, it is difficult to compute the amount of entropy produced by an irreversible process in nonequilibrium situations.

The idea of evolution in biology is associated with an increase in organization giving rise to the creation of more and more complex structures. Irreversibility in thermodynamics indicates that evolution with time implies the attainment of the most probable state, corresponding to a state of maximum disorder. The second law assures the existence of the state function, entropy, S, which depends on thermodynamic variables of the system. The entropy of an open system, which can exchange both matter and energy with its surroundings, can be changed as a result of two kinds of phenomena: (1) irreversible processes occurring within the system itself, and (2) irreversible processes such as heat and matter exchange with its surroundings. Thus, the entropy entering the open system with heat or matter as one of the parameters rather than the entropy of dissipation becomes significant. For systems which exchange matter and energy with their environment, the variation of entropy dS during a time interval t may be factored as

$$\{dS/dt\} = \{d_e S/dt\} + \{d_i S/dt\} \tag{8.1}$$

where $\{d_e S/dt\}$ denotes the entropy flow from the environment into the system, and $\{d_iS/dt\}$ is the entropy production within the system due to irreversible processes occurring.

Classical thermodynamics stipulates that isolated systems (that is, systems which do not exchange matter or energy with their surroundings) always proceed toward an equilibrium state. During the attainment of equilibrium, the entropy of the system increases in a monotonic manner and reaches a maximum value in the equilibrium state. Thus, for an isolated system, when there is no exchange of matter or energy with the system's surroundings, $\{dS/dt\}$ is positive definite. For a closed system, which can exchange energy but not material with its surroundings, $T\{d_eS\}$ equals $\{dQ/dt\}$ where Q is heat and T is temperature in absolute scale. Again, $\{dS/dt\}$ should be greater than $T^{-1}\{dQ/dt\}$. In an open system, d_eS may be positive, negative, or equal to zero, depending on the inflow into the system and the outflow of entropy from the system.

The second law requires that

$$d_iS \geq 0 \tag{8.2a}$$

$$\{d_iS/dt\} \geq 0 \tag{8.2b}$$

The equality sign applies only for equilibrium situations. For an isolated system, d_eS equals zero. Therefore, one obtains the second law for isolated systems, as one learns from classical thermodynamics.

A process tending toward equilibrium is irreversible. A change for which $\{dS/dt\}$ equals zero is reversible. All deviations from equilibrium result in a decrease in entropy. A system in equilibrium, therefore, responds to disturbances caused by fluctuations by reverting to equilibrium. Fluctuations around the equilibrium states are self-regulating. This control mechanism is inherent in all systems at or near equilibrium. The stationary state of a system, according to the thermodynamics of irreversible processes, is characterized by the principle: In the stationary state and with unchanged external parameters, the rate of entropy production in the system is constant and minimal.

The existence of order in biological systems is evident. It is insufficient to consider only orders relevant to spatial arrangements. Physical order expresses itself, not only in spatial order but also in what Frohlich (1988) denotes as *motional order*. Motional order is expressed in terms of macroscopic wave functions and refers to certain phase correlations yielding a coherence. At very low temperatures when entropy vanishes, where according to the third law of thermodynamics, the nonrandomness prevails and no disorder exists, the atoms of superfluid helium exhibit the same kind of disordered correlation as do other fluids at higher temperatures. Thus, the idea energes that phase correlations of some kind, coherence, play a decisive role in a cohesive manner in the description of biological activities. Thus, in nature, two kinds of structures may be observed to exist. The first kind, *equilibrium structures*, may be maintained without any exchange of energy or matter with the outside world. The second kind is called the *dissipative structures*, which can be maintained only by the flow of energy and/or matter from the outside world through the system.

The chaotic behavior of the dynamics of systems at the molecular level yet leads to observations suggesting coherent behavior at the macroscopic level in such systems. Prigogine and his coworkers emphasize that the physical chemistry of systems involving large deviations from thermodynamic equilibrium may be completely different from traditional physical chemistry. Nonequilibrium systems are constrained systems which involve matter and energy transport, arising from their interactions with their surroundings. States displaying spatial or temporal organization may be reached.

Large systems that are not in thermal equilibrium may exhibit properties that are not noticeable in individual particle motions, due to fluctuations which are not strictly reproducible, when nonlinear processes occur. Noise may cause phase transitions in excited nonlinear systems. Bifurcations are the most important phenomena occurring in nonlinear systems. They are responsible for a large series of different states that the system occupies for different excitations. Often bifurcating systems exhibit hysteresis.

Metabolism and the dissipation of energy play an essential role in living systems. In physical systems, a compromise is attained between the randomness caused by entropy and the order brought about by energy terms. When two liquids are mixed, diffusion processes occur, with a progressive forgetting by the system of its initial conditions. On the contrary, in biological systems, heterogeneity is the prevalent condition. The dissipative structure theory assumes that the same physical law operates in biological systems. The nature of its manifestation depends on whether the state of the system is near or far from equilibrium state.

The formation and maintenance of order at states far from equilibrium involve the cooperation of a large number of molecules. In biology, various delicate mechanisms for coordination and regulation of multiple chemical reactions of metabolism are required. Such metabolic reactions require specific enzymes as catalysts. Thus, an extremely sophisticated functional organization is mandatory. The cell performs a complex set and sequence of operations. Therefore, a definite relation between structures as expressed in space-time patterns and function must exist. Biological order should thus satisfy the requirements of being both functional and architectural.

The expression for free energy,

$$F = E - TS \tag{8.3}$$

implies that as the temperature T is decreased, contribution of the energy term E to Helmholtz free energy F becomes more dominant, thus reducing the role of the entropy S which is related to disorder. This is the basic mechanism underlying changes of state, for example, in the freezing of liquids. Thus, the complicated structure formation related to the organization of equilibrium structures is facilitated by lowering the temperature.

The second possibility of dissipative structure is maintained in a stable manner, by the flow of matter and energy through the system. When a pan of liquid is heated from below, heat passes through the liquid by the mechanism of conduction, when the gradient of temperature is below some threshold characteristic value. As the rate of heating is increased, at a certain well-defined value of the temperature gradient, the interaction of gravitation forces and thermal forces result in the appearance of convection cells, which are regular, appearing spontaneously. These correspond to a high degree of molecular organization and such stable structures become possible through the transfer of energy from thermal motion to macroscopic convection currents. This is known as *Benard instability*. For a critical value of the temperature gradient, there is an abrupt onset of convection. The stability occurs as a result of the competition between the two opposing tendencies, the stabilizing dissipative effects and the destabilizing convection effects. Such a competitive aspect is quite general in problems involving the possibility of an instability.

Dissipative effects generated by the temperature fluctuations are dominant in the range of small Raleigh numbers. The importance of velocity fluctuations appear only for higher values of the Raleigh numbers. In the region of marginal stability, the entropy production varies suddenly with the appearance of the first unstable normal mode. The new solution generated by this normal mode leads to a discontinuity in the slope of the entropy production.

The adverse temperature gradient, which is a constraint on the system, plays the same role for dissipative structures as the role of temperature for equilibrium structures. The state of the fluid in the region of thermal convection may be considered as an example of dissipative structures. Its entropy is lower than that of a system in which all the energy would be in the thermal motion. Below a certain critical value of the temperature gradient, certain fluctuations are damped and eventually become negligible. Above some critical value of the temperature gradient, fluctuations are amplified and give rise to a macroscopic current. A new molecular order appears that corresponds basically to a situation of macroscopic fluctuations being stabilized by exchanges of energy with the outside world. Most of the properties of such hydrodynamic instabilities are also exhibited by systems obeying nonlinear chemical kinetic laws.

The situation where a nonequilibrium state may be the source of order is also illustrated by the example of thermal diffusion. When a thermal gradient is applied to a binary mixture of gases, it is experimentally observed that one of the components concentrates near the hot wall, while the other concentrates at the colder wall. Such separation of a uniform mixture into its components, regarded as not possible without the expenditure of energy according to equilibrium thermodynamics, becomes feasible under the nonequilibrium situation.

Chemical instabilities occur when there exist specific nonlinear kinds of kinetics, with a minimum level of dissipation. Such instabilities lead to spatial organizations, which is termed *self-organization* of the system from the point of view of its spatial order and its function. The emergence of spatial order and the transmission of information are also the central problem of morphogenesis and more generally in the problems of cell differentiation.

In equilibrium thermodynamics, it is not necessary to show that stability conditions are satisfied, since the free energy just needs to be a minimum. Similarly, in the linear range of nonequilibrium thermodynamics, stability conditions are automatically satisfied. The occurrence of dissipative structures is associated with the existence of instabilities at which one branch of the kinetic equations becomes unstable and is replaced by a new branch. Thus, dissipative structures can exist only in nonlinear systems for which more than one solution of the kinetic equations exist.

In equilibrium thermodynamics, the relation between extensive and intensive variables is given by a generalized equation of state. In nonequilibrium thermodynamics, the relation connecting thermodynamic force and flow variations are given by the phenomenological relations. The phenomenological relations are not independent of each other and satisfy the condition of integrability. (The condition of integrability implies that the generalized thermodynamic potential is independent of the path of integration. It depends only on the initial and final states of the system.)

The thermodynamics of irreversible processes requires that the thermodynamic branch be unique within the linear range. Instabilities can occur only outside the range of linear thermodynamics of irreversible processes. Dissipative structures will occur only at a *finite distance* away from the thermodynamic equilibrium, as the stability of the thermodynamic solution extends over at least some nonequilibrium region. (Finite distance in this case appears to mean that the value of a *constraint* is above a threshold value). The appearance of the dissipative structures may be considered as similar to the appearance of an equilibrium structure in the phase space.

A new structure is always the result of an instability. It originates from fluctuations. At the point of formation of a new structure, fluctuations are amplified. Generally, one finds large deviations from linearity between chemical flows and forces at states far from equilibrium. These deviations are of particular importance in the case of autocatalytic reactions. One finds concentration ranges where flows and fluxes have opposite signs in their dependence on the reaction variable. This negative force-flux characteristic is essential for several kinds of phenomena like oscillations of concentrations or spatial ordering to occur.

NONEQUILIBRIUM THERMODYNAMICS _____

For the development of nonequilibrium thermodynamics, one assumes that for states of the system far from equilibrium conditions, the Gibbs' relation (Equation 8.4) for the entropy is valid.

$$dS = d_eS + d_iS$$

$$= \{dE/T\} + (p/T)\,dV - \sum_k \{\mu_k/T\}\,dn_k \tag{8.4}$$

where E is the internal energy, V is the volume of the system, p is pressure, and T is temperature. n_k is the number of moles of the kth component and μ_k is the molar Gibbs' free energy or chemical potential of component k. The validity of Equation 7.4 can be demonstrated only for near-equilibrium situations.

The progress variable, ξ, or the extent of a chemical reaction is defined by the relation

$$d\xi = \{1/\nu_k\}\,dn_k \tag{8.5}$$

for all species k. ν_k is the stoichiometric coefficient of species k, in the chemical reaction occurring. The velocity of the chemical reaction v equals $\{d\xi/dt\}$. In near-equilibrium situations, the rate of entropy production equals the product of velocity of the chemical reaction and its affinity divided by the temperature.

In the development of nonequilibrium thermodynamics, emphasis is placed on the macroscopic description of states on the basis of the entropy production. The form of entropy production is derived from the mass-balance equations and conservation relations for energy and momentum. The assumption of local equilibrium is needed to extend and formulate thermodynamics for the nonequilibrium situation. Local entropy will be expressed in terms of the same relations involving the same state variables as in the equilibrium case.

The entropy production per unit volume per unit time, σ, is defined by the relation

$$P = \{d_iS/dt\} = \int \sigma\,dV; \qquad \sigma > 0 \tag{8.6}$$

$$\sigma = \sum_k J_k X_k \tag{8.7}$$

In the linear range, one can express forces as linear functions of fluxes with the resultant definitions of phenomenological coefficients. The range of validity of the bilinear form of entropy production (Equation 8.7) defines the scope of the linear thermodynamics of irreversible processes. In systems with chemical reactions, this equation is valid only for an extremely small deviation from equilibrium.

In a system at equilibrium, for all conceivable processes, the various forces and fluxes individually equal zero. Consider now a system in which a constant temperature difference is maintained between two surfaces, and matter is prohibited from entering or leaving the system. There are then two forces, X_E and X_M, defined respectively by the difference in temperature and concentrations and two conjugate fluxes, that of energy J_E and that of matter J_M. After a sufficiently long time, the system would have reached a state in which the redistribution of matter would have removed the difference in chemical potential, while the transport of energy between the boundary surfaces will be in a stationary state. Entropy increases continuously in a nonequilibrium state. In the linear range, one has

$$\sigma = J_E X_E + J_M X_M > 0 \qquad (8.8)$$

$$J_E = L_{EE} X_E + L_{EM} X_M \qquad (8.9)$$

Substitution of Equation 8.9 in Equation 8.8 and combining with Equation 8.7 yields the quadratic form

$$\sigma = L_{EE} X_E^2 + 2 L_{ME} X_E X_M + L_{MM} X_M^2 \qquad (8.10)$$

Differentiation of Equation 8.10 with respect to X_M, at constant X_E, one has the partial derivative

$$\{\partial / \partial X_M\} \sigma = 2\{L_{ME} X_E + L_{MM} X_M\} = 2 J_M = 0$$

Since matter flux J_M equals zero, one has the extremum principle, known as the *minimum entropy production*,

$$\{\partial / \partial X_M\} \sigma = 0$$
$$\qquad (8.11)$$
$$J_M = 0$$

The minimum entropy production principle leads to the conclusion that σ is a nonequilibrium state function playing the same role as the thermodynamic potentials in the equilibrium theory. The minimum property assures the stability of the steady state. The theorem provides an evolutionary criterion, since it implies that a physical system will necessarily evolve to the stationary nonequilibrium state, starting from an arbitrary state close to it. This statement can be expressed as

$$\{\partial \sigma / \partial t\} \, dV < 0 \qquad (8.12)$$

By its very stability, the linear domain is an extrapolation of the equilibrium region and implies a monotonic approach to a single steady state, once the equilibrium itself is stable. Therefore, one has to rule out the possibility of the formation of new, ordered structures in the linear range.

The validity of Equation 8.12 is limited to the linear range. For the nonlinear range, the entropy production is factored into two parts:

$$d\sigma = d_J\sigma + d_X\sigma$$

$$d_J\sigma = \sum_i X_i \, dJ_i \tag{8.13}$$

$$d_X\sigma = \sum_i J_i \, dX_i$$

It is shown that for purely dissipative systems, with time-independent boundary conditions, for which the equilibrium state is stable, it may be shown that

$$d_X\sigma \, dV \leqslant 0 \tag{8.14}$$

The equality sign applies for steady state only. Equation 8.14 is known as the *evolutionary criterion*. Thus, in the nonlinear domain, the entropy production needs to be factored in order to obtain the general inequality.

It should be recognized that $d_X\sigma$ is not a total differential of some state function. It is only in the limit of validity of the linear relations for irreversible processes, that $d_X\sigma$ becomes a differential of the state function. In the range of the local equilibrium assumption, the effects which are responsible for this property are due to chemical reactions. The validity of Equation 8.7 in the general range of nonlinear effects (namely, the factorization of the total entropy production as an explicit sum of various terms) needs to be demonstrated. Not being a total differential, $d_X\sigma$, implies that no true variational principle exists far from equilibrium.

---

† Zotina and Zotin (1982) suggest that when the formation of heat in a system which is far from equilibrium is high, not all this heat will have time to leave the system. Thus, a portion of this can be utilized by the system to do useful work. On this basis they propose that the dissipative function of a system far from equilibrium can be separated into two parts:

$$DF = DF_1 + DF_2 = \sum_k J_k X_k$$

$$DF = (T/V)\{d_i S/dt\}$$

$$DF_1 = (T/V)\{d_i S_i/dt\}$$

$$DF_2 = (T/V)\{d_i S_2/dt\}$$

where T is the temperature in Kelvin scale, V is the volume of the system, and $\{d_i S/dt\}$ is the rate of entropy production. DF is the specific dissipation function, DF_1 is the external dissipation function, DF_2 is the bound dissipation function. J_k and X_k are the conjugate force and flux of kind k. The external dissipation function equals the heat production intensity of the system. In purely chemical systems, DF_2 depends on the temperature, the size of the system, and the entropy change being related to the extent of the completion of the reaction.

In equilibrium situations, perturbing fluctuations are usually followed by a response of the system that brings back the system to its original unperturbed state. This is stated by the Boltzmann relation (also known as the Einstein relation) that the probability of fluctuations in an isolated system may be expressed by the formula:

$$\text{Probability of fluctuations} = \exp\{\Delta S / k_B\} \tag{8.15}$$

where ΔS is the change in entropy starting from equilibrium, $(\Delta S < 0)$, associated with fluctuations, and k_B is the Boltzmann constant. Since the entropy of an isolated system cannot decrease, the second law of thermodynamics has been a central guiding principle of all fluctuation theories for near-equilibrium states. One can show easily that the average deviation from equilibrium value should always decay. Thus, the second law of thermodynamics guarantees the stability of the thermal equilibrium state. It appears that there is no unique principle for the nonequilibrium fluctuations corresponding to the second law of thermodynamics.

In order to achieve a thermodynamic description of irreversible processes, it becomes essential to assume that the fluctuating force has a short time correlation. Onsager assumed that the mean fluctuation obeyed a linear equation of motion, if the mean fluctuations are small. The regression of the fluctuation in a state near equilibrium was then investigated. In the study of Brownian motion, the importance of a fluctuating force, which is the rest part of the equation of motion for fluctuation, was recognized by Langevian. The decay constant of the mean fluctuation in a state near equilibrium is closely related to the fluctuating force through a simple relation known as the *Fluctuation-dissipation theorem*.

The thermodynamic description of irreversible processes from statistical mechanical arguments is to assume that the fluctuating force has a short time correlation. Two questions arise in connection with fluctuations in nonequilibrium states.

1. How can one derive a general law governing fluctuations in a state far from equilibrium?
2. What is the practical role of a reference nonequilibrium state?

It is postualted that Equation 8.15 is valid for nonequilibrium fluctuations as well.

The second law of thermodynamics for a closed system at uniform pressure p and temperature T can be expressed as

$$T \, d_i S = T \, dS - dE - p \, dV > 0 \tag{8.16}$$

This equation leads to the stability criterion for the thermodynamic equilibrium. If no perturbation arising with equilibrium can satisfy the inequality of Equation 8.16, the system has to stay in equilibrium.

Utilizing the symbol δ to denote small but otherwise arbitrary changes, the equilibrium stability criterion becomes

$$\delta E + p\,\delta V - T\,\delta S > 0 \tag{8.17}$$

The change in signs between Equations 8.17 and 8.16 should be noted. The internal energy E is a minimum for stable equilibrium. For systems at constant energy E and volume V, one has the stability condition (for isolated systems),

$$\delta S < 0; \qquad E,\ V \text{ constants} \tag{8.18}$$

The relation between the thermodynamic stability and the entropy-balance equation leads to a new formulation of the stability theory of equilibrum. One obtains

$$\sum_k J_k X_k\,dV = \{\partial S/\partial t\} + P[S] \geq 0 \tag{8.19}$$

where $P[S]$ denotes the total entropy production of the whole system. The left-hand side of Equation 8.19 is the entropy production term, which is a quantity of second order with respect to deviations from equilibrium. The entropy production term is the integrated sum of the product of the conjugate fluxes and forces.

In the range of nonequilibrium situations, where linear laws are valid, the entropy production term consists of two uncoupled terms, namely, entropy production due to chemical reactions and entropy production due to matter flows.

$$\sigma_{\text{Total}} = \sigma_{\text{Chemical}} + \sigma_{\text{Fluxes}} \tag{8.20}$$

with separate factored contributions,

$$\sigma_{\text{Chemical}} = \sum_r W_{ir}\{\mathscr{A}_r/T\} > 0$$
$$\sigma_{\text{Fluxes}} = -\sum_k J_k\{\nabla\mu_k/T\} \tag{8.21}$$

where $\mathscr{A}_r$ is the affinity of the chemical reaction r.

The right-hand side of Equation 8.19 can be factored into terms of first and second order. Expanding the entropy S about its equilibrium value S_{eq}, and retaining terms only up to second order,

$$S_{\text{eq}} = S + \{\delta S\}_{\text{eq}} + (1/2)\{\delta^2 S\}_{\text{eq}} \tag{8.22}$$

Since S_{eq} is independent of time, one has

$$\{\partial/\partial t\}S = \{\partial/\partial t\}\{\delta S\}_{\text{eq}} + (1/2)\{\partial/\partial t\}\{\delta^2 S\}_{\text{eq}} \tag{8.23}$$

The entropy flow term $P[S]$ can also be factored into two terms similarly. Use of the entropy-balance equation enables one to identify the first-order and second-order terms separately. Such separation of the entropy-balance equation into two separate relations is not always valid. The entropy-balance equation leads to the stability condition

$$(1/2)\{\partial/\partial t\}\{\delta^2 S\}_{eq} = P[S] \geqslant 0 \tag{8.24}$$

Equation 8.24 yields an evolution criterion for states near equilibrium. It relates the time derivative of the curvature, $\{\delta^2 S\}_{eq}$ to the entropy production, that is, to the irreversible processes inside the system.

The thermodynamic theory of stability leads to the inequality,

$$\delta^2 S_{eq} < 0 \tag{8.25}$$

and

$$\{\partial/\partial t\} \, \delta^2 S_{eq} = 2P[S] > 0 \tag{8.26}$$

$\delta^2 S_{eq}$ is the second-order variation of the equilibrium entropy. The stability depends only on the sign of the curvature of $\{\delta^2 S\}_{eq}$ evaluated at the equilibrium state. This is similar to determining in calculus whether the extremum is a maximum or a minimum. The problem of regression of fluctuations leads to the study of the time evolution of $\delta^2 S_{eq}$. Macroscopic nonequilibrium evolution considerations lead to the validity of the inequalities of Equation 8.25.

In a system in which transport processes and chemical reactions are occurring, the state variables of the system satisfy the balance equations

$$\{\partial C_i/\partial t\} = -\text{divergence}\, J_i + \sum_r n_{ir} W_r \tag{8.27}$$

where J_i are matter fluxes of species i, expressed in moles per unit area per unit time. W_r is the rate of the chemical reaction r, which can be related to the concentration variables, C_i. W_r in most situations are nonlinear polynomial functions of the concentration variables. n_{ir} is the stoichiometric coefficient of species i while participating in the chemical reaction r.

Entropy production is expressed as

$$P = \int \sigma \, dV = \int dV \left\{ -\sum_i J_i \cdot [\nabla \mu_i / T] + \sum_r W_r [\mathscr{A}_r / T] \right\}$$

$$= \int dV \sum_k J_k X_k \tag{8.28}$$

Both fluxes J_i and the rate of reactions W_r are related to the concentration values C_j through the appropriate set of nonlinear phenomenological relations.

In a system with chemical reactions, the state of the system is described by nonlinear functions of the variables involving concentrations and temperature, as well as position variables if the system is inhomogeneous. Such a system is a dynamical system where the concentrations may be regarded as a set of state variables for the system. The factors affecting the values of these concentrations at a specified time are the chemical reactions occurring in the system. The rate of the change of concentrations with time can be formally expressed as

$$\{\partial C_i / \partial t\} = F_i[C_i, C_{i+1}, \ldots, C_n]$$

$$i = 1, 2, \ldots, n$$

(8.29)

A number of restrictions can be imposed on the form of the functions F_i in consequence of our knowledge about the chemical reaction mechanisms and about the various participating species. A basic restriction also arises from the law of mass action.

Such a chemically reacting mixture described by nonlinear equations in general can have more than one solution, even when the boundary conditions are taken into account. The *thermodynamic branch* is the solution corresponding to the equilibrium situations, such as maximization of the entropy for isolated systems or the minimization of Helmholtz free energy for systems at a given temperature and volume. If the constraints are varied to force the system to be further and further away from equilibrium, the nonequilibrium thermodynamics permits one to formulate a sufficient condition for the stability of the thermodynamic branch. When such stability conditions are not satisfied, the thermodynamic branch may become unstable. The system may evolve toward a new structure involving a coherent behavior.

Beyond the instability of the thermodynamic branch, one has a new type of organization relating the coherent space-time behavior to the dynamical processes inside the system. Only if appropriate feedback conditions are satisfied, the thermodynamic branch can become unstable at a sufficient distance away from equilibrium. The new structures that appear in this manner can be maintained in a far-from–equilibrium situation only through sufficient flow of energy and matter. It should be recalled that the stationary states, where linear laws are valid, are also maintained and the system is prevented from attaining an equilibrium state by just such maintenance of flow of energy and matter.

Stability has been emphasized and discussed from various points of view in the preceding chapters. Stability can be formulated and discussed in terms of the thermodynamic potentials such as free energy or entropy. In equilibrium thermodynamics, once we know that a system is in a state of minimum free energy, we may conclude that it is stable. Even if fluctuations will cause a momentary deviation, the system would respond by reducing its free energy until it reaches the equilibrium value. In general, in non-equilibrium situations, no such potential whose value will characterize the state of the system exists. The stability considerations become important for this reason. The problem of emergence of new patterns is to study possible branching of solutions that may arise under certain conditions. The multiplicity of solutions that arises in the nonequilibrium nonlinear systems correspond to a gradual acquisition of autonomy from the environment.

In dissipative structure theory, a new structure is asserted to be always the result of an instability, the origin of which is fluctuations. Whereas fluctuations are normally followed by the response of the system to bring back the system to the unperturbed equilibrium state, at the formation of these "new structures", fluctuations are amplified. This is the basic idea of the classical stability theory of fluid dynamics in terms of normal-mode analysis (Chandrasekhar, 1961).

The problem of regression of fluctuations, or equivalently the problem of the validity of the moderation principle, leads to the study of the time evolution of entropy. Whenever, the excess entropy production has a positive sign, the system is stable. Near equilibrium, this condition is always fulfilled. Corresponding to the transition between stability and instability, the excess entropy production vanishes. The stability problem of finite amplitude wave propagation in ideal fluids leads to the conclusion that the excess entropy production appears as either a positive definite or a negative definite function.

For systems with chemical reactions in a closed system, the kinetic condition of stability may be started as progress of a reaction, defined as the deviation of a reaction from its equilibrium, should tend to vanish as time increases. This condition is somewhat self-evident, since in order for chemical reactions to reach equilibrium, the deviation should decrease as time increases. Assuming that linear phenomenological relations can be defined in the neighborhood of the equilibrium state, the rate equations admit exponential solutions, which are termed *normal-mode solutions*. In order to ensure stability, evidently the exponential coefficients which are the real parts of the eigenvalues of the appropriate matrix should be negative definite. Naturally, affinities should vanish at equilibrium. The imaginary parts of the eigenvalues, which will be complex in their most general form, lead to oscillating chemical disturbances. These should also vanish at equilibrium. Therefore, oscillating chemical reaction perturbations do not exist around an equilibrium state. Glansdorf and Prigogine (1971) show that all the imaginary parts of the eigenvalues vanish identically.

Based on these, one concludes that there are both thermodynamic and hydrodynamic stability considerations which play an important role in determining the stability states of a specified system. In linear stability theory of stationary states which are subject to small perturbations, one has to show only that the real parts of eigenvalues are negative definite for each normal mode. This kinetic criterion should be supplemented by a thermodynamic criterion, resulting in a sufficient condition for the stability of the stationary states. Similar demonstration in general for a nonequilibrium state is difficult. The foregoing presentation completes our qualitative presentation of the theory of nonequilibrium, nonlinear thermodynamics. In the following we list certain questions that may arise regarding the theory of dissipative structures.

1. What happens beyond the instability of the thermodynamic branch?
2. What are the kinds of coherent behavior and how are they related to the molecular mechanisms involved, as well as to the constraints acting on the system?
3. What is the kinetics of growth of the dissipative structures, and how can fluctuation theory be applied to these nonlinear, far-from-equilibrium situations?
4. When do we know that a system is at a state sufficiently far from equilibrium and what do we mean by this? How far is far? Is this a qualitative or a quantitative statement?†
5. When does the assumption of local equilibrium fail to be valid? Evidently, for thermodynamic concepts to be valid, the volume element should be large enough to contain a sufficient number of particles, but should be small enough for local homogeneity to prevail at least for short ranges.
6. What are the situations to which one may beneficially apply the concept of dissipative structures and order through fluctuations?
7. Can one factor the entropy production as the sum of the products of conjugate fluxes and conjugate forces for all nonlinear far-from-equilibrium situations as done in Equation 8.7?
8. The processes occurring in living systems may not subsist when the systems die. Death is regarded as an equilibrium state. The structure

---

† Schuster (1983) assumes the existence of a parameter h, which could be a concentration difference or another measurable quantity, which can be controlled from outside and which describes the deviation from equilibrium. Equilibrium is described by the value of this parameter h equaling zero. For small values of h, one has steady states with properties similar to the equilibrium state. Provided the kinetics has a proper nonlinearity, the thermodynamic branch becomes unstable at a certain critical value. The location at which the thermodynamic state becomes unstable is denoted by h assuming a critical value h^*. This critical point is called the bifurcation at which the thermodynamic branch becomes unstable and a new state becomes stable at least for a certain range of parameters. He discusses the example of the reaction $A + B \rightarrow 2A$, since this shows for a particular range of concentrations, far off equilibrium, a negative force-flow characteristic. Schuster concludes that such a negative force-flux characteristic implies that a decrease in the chemical force is accompanied by an increase in the (velocity) chemical flow.

and organization as well as the compartments with components persist in death until decomposition starts. To what extent are these structures of molecular organization compatible with equilibrium structures and to what extent are these nonequilibrium dissipative kinds of structures?†

9. How does one construct the solutions that occur when the thermodynamic branch becomes unstable?

10. Can the problem of regulation and control in biology be solved with the concepts and theory of dissipative structures? The paper of Lavenda (1972) and the monograph of Nicolis and Prigogine (1977) imply that this can be done. Are not the solutions to the specified problems of regulation and control just the solutions of appropriate nonlinear differential equations?

11. Can one give separate meaning to the term *active transport,* as distinct from the influence of chemical reactions on fluxes, under nonequilibrium, quasi-stationary-state conditions? It is stated that beyond the stability of the thermodynamic branch, a decrease in force results in an increase in flux. How does this negative force-flux relation range arise for matter fluxes and matter forces? Are we postulating bounds for such processes to occur, or are we specifying the mechanisms by which processes such as active transport occur?

It is certainly tempting to describe biological structures as open chemical systems, operating beyond the stability region of the thermodynamic branch. In order to describe the essential features of life or of living systems, such as replication, much more information is needed. By means of the cycles of chemical reactions, organisms appear to be able to maintain their coherence. The manner in which chemical reactions influence fluxes of species in biological membrane systems is not explained by the dissipative structure theory. We conclude this chapter with the presentation of an analysis of the presence of oscillations in chemical reactions, in the context of dissipative structure theory.

---

† It is instructive to recall the discussion that took place in a meeting in 1969. We quote Mazur's remarks regarding the concept of dissipative structures (Mazur, 1969, pp. 196–97).

In Progogine's development, the equilibrium state is the uniform state; while in Katchalsky's treatment there is basically already a structure; one or more membranes. As it stands, Katchalsky takes the existence of the membrane for granted. Even at equilibrium, the membrane is still there. Equilibrium simply means that on both sides of the membrane you have the same conditions. Prigogine's approach on the other hand provides the possibility to create an inhomogeneity that is perhaps of a membrane.

Are these two pictures complementary or contradictory?

OSCILLATIONS _____

The possibility of the existence of oscillating chemical reactions and their relevance to certain biological phenomena have been of interest for the last fifty years. In the Lotka-Voltera model, oscillations are realized in the thermodynamic branch. Consider the following reaction scheme:

$$A + X \underset{k_1^\star}{\overset{k_1}{\rightleftharpoons}} 2X$$

$$X + Y \underset{k_2^\star}{\overset{k_2}{\rightleftharpoons}} 2Y \qquad (8.30)$$

$$Y \underset{k_3^\star}{\overset{k_3}{\rightleftharpoons}} E$$

When the concentrations of the initial and final products, A and E, are maintained constant, so that only two independent variables X and Y remain, the thermodynamic state of the system is characterized by the overall affinity. This corresponds to the conversion of A into E:

$$\mathscr{A} = \mathscr{A}_1 + \mathscr{A}_2 + \mathscr{A}_3$$
$$= RT \ln[\{(k_1^\star k_2^\star k_3^\star)/(k_1 k_2 k_3)\}(A/E)] \qquad (8.31)$$

At equilibrium, one has

$$(A/E)_{eq} = \{(k_1 k_2 k_3)/(k_1^\star k_2^\star k_3^\star)\}$$

$$X_{eq} = [k_1/k_1^\star]A \qquad (8.32)$$

$$Y_{eq} = \{[k_1 k_2]/[k_1^\star k_2^\star]\}A$$

The behavior with time of the preceding system of reactions can be studied for the two extreme cases, one in which the affinity is very small and the other where the affinity is very large.

If the ratio of concentrations (A/E) is only slightly different from the equilibrium value, linear laws between affinities and reaction rates may be written. In this situation, the excess entropy production is positive. Thus, an arbitrary fluctuation regresses in an aperiodic manner to steady state.

When the reverse reactions are very small such that the reactions are unidirectional, the overall affinity tends to be infinity. The kinetic equations then are

$$\{dX/dt\} = k_1 AX - k_2 XY$$
$$\{dY/dt\} = k_2 XY - k_3 Y \qquad (8.33)$$

In this situation, a single nonvanishing steady-state solution exists:

$$X_o = \{k_3/k_2\}$$
$$Y_o = \{k_1/k_2\}A \qquad (8.34)$$

These equations are similar to the predator–prey model of Volterra. The stability property of Equations 8.33 are studied by the normal-mode analysis. In the neighborhood of a stationary state, the time dependence of X and Y may be written as

$$X(t) = X_o + x \exp(wt)$$
$$Y(t) = Y_o + Y \exp(wt)$$

(8.35)

with the conditions

$$\{x/X_o\} \ll 1$$
$$\{y/Y_o\} \ll 1$$

(8.36)

Substitution of Equation (8.35) in Equation (8.34) and neglect of higher-order terms yield a set of linearized equations:

$$w \, \delta X + k_3 \, \delta Y = 0$$
$$-k_1 A \, \delta X + w \, \delta Y = 0$$

(8.37)

yielding a dispersion relation,

$$w^2 + k_1 k_3 A = 0$$

(8.38)

Therefore, small fluctuations around the steady state are now periodic, with the frequency,

$$w_i = \pm\{k_1 k_3 A\}^{(1/2)}$$
$$w_r = 0$$

(8.39)

Thus, the perturbed system will remain in the neighborhood of the steady state. The thermodynamic stability conditions indicate that $\delta_X P$ vanishes around the steady state. Therefore, $\delta^2 S$ appears as a constant of motion for arbitrary disturbances. Since only the real part of w vanishes, the perturbed state cannot be interpreted as another steady state very close to the original state. Lefever and his coworkers (1967) show that the Lotka-Volterra scheme of reaction provides a model for sustained oscillations in a chemical system, so long as the value of the overall affinity, $\mathscr{A}$, is very large.

The main interest of the Lotka-Volterra model rests in the fact that perturbations at finite distances from the steady state are also periodic with time. Therefore, for the Lotka-Volterra system, there exists a continuous spectrum of frequencies due to the infinite number of possible cycles depending on the initial cycle. Each cycle appears as a state of marginal stability where even a small perturbation is sufficient to change the motion of the system to a new cycle corresponding to a different frequency. In the Lotka-Volterra system, because there is no mechanism for the decay of fluctuations, there exists no average orbit in the neighborhood of which the system is maintained. Oscillations of this kind cannot be expected to lead to reproducible observations as far as the amplitude and frequencies of oscillations are concerned. Only the orbits infinitesimally close to the steady state may be considered stable. The situation where the excess entropy production first vanishes and then changes its sign for the finite value of the overall affinity exhibits that the thermodynamic branch becomes unstable. For such instability, it is essential that an autocatalytic step is included.

The differential equations which arise in connection with oscillating reactions are in general nonlinear second-order systems of equations. It is very doubtful whether Volterra equations represent a set of chemical reactions which may account for the oscillations occurring in biology for the reasons presented earlier. Periodic phenomena play a role in the early stages of aggregation and in the differentiation of certain amoebae belonging to the family of slime molds.

If one has a set of first-order chemical reactions of the form

$$\{dC_k/dt\} = -\left\{\sum_{i \neq k} W_{ki}\right\}C_k + \sum_{i \neq k} W_{ik}C_i$$

$$= \sum_i W_{ik}C_i \qquad (8.40)$$

defining $W_{kk} = -\{\sum_{i \neq k} W_{ki}\}$. Written in a matrix form, the solution of the resulting set of equations is that C_i is a linear combination of the exponentials, $\exp(nt)$, where the possible values of the eigenvalues, n, are determined from a secular equation. The stability of the system is determined by the properties of the eigenvalues.

In order that the equilibrium be thermodynamically stable, it is necessary that the eigenvalues have a real negative part. Since it is in general not possible to solve the secular equation for the eigenvalues, one is faced with the problem of how to determine the sign of the real part of the eigenvalues without actually solving the set of equations (Segal, 1976).

BIOSYNTHETIC CONTROL PROCESSES

Regulation of protein synthesis is the result of a combination of structural genetic factors and dynamic chemical processes (Levenda, 1972). It is currently believed that control can be exercised at the initial stage of the biosynthetic pathway, through the indirect influence of the regulator genes. The regulator gene synthesizes a protein (repressor), which under certain conditions prevents the process of transcription by binding with a functional site on the structural gene, known as the *operator*. The activity of the repressor is a function of its conformational state. Changes in configuration are induced by the binding of certain small molecules called the *inducers*. If the inducers are metabolites of a certain biosynthetic pathway, a feedback mechanism therefore exists. Information about the rates of synthesis at various stages of the biosynthetic pathway is transmitted from one end to another. When bound to the repressor, the inducer inhibits the binding of the repressor to the operator so that protein synthesis proceeds. The more protein is synthesized, the more metabolite is formed.

The cyclic nature of the chemical events leads to an increase in the rate of synthesis. The rate is assumed to be a function of the metabolite concentration. Each stage of the biosynthetic pathway involves numerous and complicated chemical and physical processes. The conformational transitions of the repressor are a function of the metabolite concentration. Thus, the underlying mechanism is the dependency of specific conformations of the repressor upon the concentration of the metabolite. A relationship between the molecular configuration of the repressor and the kinetic processes can lead to a control mechanism based on multiple stationary-state transitions.

Levenda (1972) argues that if one assumes that more than one stationary-state regime exists in the biosynthetic pathway, the nonequilibrium evolution of the system would then provide a basis for integrating the action of different segments of the pathway and environmental factors. In this manner, control and regulation of biosynthesis could be shown to be the result of the stability characteristics of specific nonequilibrium states. Nonequilibrium constraints can be parametrized in terms of substrate concentration.

Control is achieved through the interaction of two or more independent, dynamical processes, whose characteristic times of evolution are of the same order of magnitude. Control can be considered as the outcome of a competition between the interacting processes. Each stage of the biosynthetic pathway involves numerous and complicated physical and chemical processes. A reorganization of the entire biochemical process occurs as a result of an unstable transition. Stability appears as a guiding principle which serves to unify the levels of biological organization and direct the dynamical evolution of all physical processes. The build up of the metabolite concentration, through progressive increase in chemical rate as the system evolves further from equilibrium, creates a feed-back phenomenon. An unhealthy growth of biosynthetic materials can induce a qualitative change in the functioning of the system of chemical reactions at a critical point. Instability serves the purpose of integrating the different levels of molecular and dynamic control of biosynthesis. The characterization of the stability properties of stationary states is through the extremum properties of a potential function.

Levenda shows that provided all lower order phenomena are stable, control resides in the most complex, interacting process. The energy generated by lower order phenomena is utilized by the more complex system processes. Thus, when one process is completed or the energy source has been expended, a change in the nonequilibrium constraints induces instability, thereby transferring the control process to a different level of biological organization.

REFERENCES

CHANDRASEKHAR, S., *Hydrodynamic and Hydromagnetic Stability.* Oxford: Clarendon Press, 1961.

FROHLICH, H., *Biological Coherence and Response to External Stimuli.* Berlin: Springer-Verlag, 1988.

GLANSDORF, P., and I. PRIGOGINE, *Thermodynamic Theory of Structure, Stability and Fluctuations.* New York: Wiley-Interscience, 1971.

LEFEVER, R., G. NICOLIS, and I. PRIGOGINE, *J. Chem. Phys.,* vol. 47, (1967), p. 1045.

LEVENDA, B.H., *Quart. Rev. Biophysics,* vol. 5, (1972), p. 429.

MAZUR, P., in *Theoretical Physics and Biology,* ed. M. Marcois. Amsterdam: North-Holland Publishing Co., (1969), pp. 196–197.

NICOLIS, G., and I. PRIGOGINE, *Self-Organization in Nonequilibrium Systems.* London: John Wiley and Sons, 1977.

ONSAGER, L., *Phys. Rev.,* vol. 37, (1931), p. 405; vol. 38, p. 2265.

PRIGOGINE, I., and G. NICOLIS, *Quart. Rev. Biophys.,* vol. 4, (1971) p. 107.

SCHUSTER, P., *Biophysics,* p. 330, eds. Hoppe, Lohmann, Markl, and Ziegler. Berlin: Springer-Verlag, 1983.

VAIDHYANATHAN, V.S., in *Molecular and Biophysics of Living Systems,* ed. R. K. Mishra. Dordrecht, Netherlands: Kluwer Academic Publishers, 1990.

ZOTINA, A.I., and A.I. ZOTIN, in *Thermodynamics and Kinetics of Biological Processes,* p. 423, eds. Lamprecht and Zotin. Berlin: De Gruyter Publishing Company, 1982.

chapter 9

Rhythmic Phenomena in Biology

INTRODUCTORY REMARKS

Oscillations of one kind or other are known to exist in biological systems. Biorhythms are part of our physiological inheritance. Though the existence of biological rhythms has been known for a long time, the existence of biological oscillators has been recognized only recently. In spite of the fact that it is known that biological rhythms are due to external periodic effects like daily changes of light and dark or similar changes in temperature, the mechanism responsible for the rhythms still remains a mystery.

It is generally believed that living systems have internal clocks. Under normal environmental conditions, physiological oscillations are entrained by external forces associated with the sunrise–sunset cycles. Many observations suggest that spontaneous physiological oscillations can often persist, with a slight deviation from 24 hours, even in the absence of external driving forces. This leads to the concept that some autonomous physiological oscillators of the limit-cycle kind are built in each living organism.

The basis of biological rhythm on earth is probably the lunar cycle. Lives of barnacles and sea urchins are governed by the low and high tides of the sea. For some organisms, the moon is probably more important than sun, since their daily life is dominated considerably by the rise and fall of the tides than alternation of day and night.

There are a large number of biological systems where the oscillatory character plays a major role in the understanding of their functional structure. The existence of oscillatory phenomena in complex biochemical sequences of reactions is a function of the existence of the servomechanisms. Oscillatory phenomena are expressions of nonlinear kinetics in a system. When open systems attain steady state, they may behave in various ways. The concentrations of a substance may approach a steady-state value in a monotonic manner, or this can be attained through rhythmically changing concentrations. The third possibility is an exhibition of some kind of persistent, oscillatory behavior around the steady state. It is appropriate to mention at this stage that linear systems which are conservative exhibit periodicity. Linear systems which are not conservative exhibit no periodicity. Nonlinear systems are capable of exhibiting time-ordered behavior. Any perturbation away from the stationary-state solution leads to sustained oscillation.

Nonlinearity is at the heart of the genesis of oscillations. Such nonlinearities are the result of autocatalytic or feedback phenomenon in systems where components are extensively coupled. At the physiological level, supercellular oscillations provide the basic property of entrainment, which is the ability to synchronize to an external periodic action. At the genetic level, oscillations may provide positional information, and when coupled with diffusion processes, they may provide signals propagated in time and space during morphogenesis. Enhanced sensitivity to environmental perturbations may be important at the metabolic level. Periodicity in the metabolic pathway may occur as a response to changing microenvironmental conditions, thus providing constant evaluation of local hydrogen-ion concentration or ionic strength.

Any level in the hierarchic organization of biology displays oscillatory states, which might be useful for many physiological tasks, such as control of processes in time and space. The possibility of oscillatory flip-flop between the activation and inactivation of reversible biochemical pathways should be recognized. The primary molecular source of an oscillatory state may be traced down to the function of a single enzymic oscillator or the coupling between two autonomously oscillating systems. There are periodic phenomena which are simply the result of a hysteretic response of an enzyme conformation toward a change of controlling ligands. Kinetic delay of the production or consumption of the chemical species in a complex enzymic reaction cycle can also produce oscillations. The conversion rates of enzymes, expressed by the rate laws, are observed to be most important for the appearance of oscillations. Metabolic regulation, however, can seldom be reduced to a single enzymic step. Interaction between two or more such oscillatory states can produce complex temporal patterns such as chaos, as exemplified by the glycolytic system.

Though oscillation can occur in a feedback control system, all oscillators need not involve feedback control. The simple harmonic oscillator does not exhibit self-sustained oscillations. Thus, it is not a system analogous to biological oscillators, where self-sustained oscillations are abundant. Quite a few biological oscillators are of the relaxation oscillator kind.

The simplest example of a *relaxation oscillator* is a dripping faucet, and the most obvious biological example is the pacemaker cell. The pacemaker action depends on some parameter of the system reaching a threshold value, at which point some property of the system changes. The threshold is the membrane potential at which the permeability property changes for the case of pacemaker cells.

It is known that the heart rate is affected by the respiratory system. The basic heart rate is produced by the spontaneous depolarization of sino-atrial nodal cells. In the von Euler model, the inspiratory switch off mechanism is comprised of three neuronal pools and its oscillator behavior can be simulated. The blood-pressure control can be considered in terms of a self-oscillatory system with a simplified dynamic structure. The influence of the respiratory system is seen in the changes of heart rate, while the influence of the blood-pressure control system on respiration is through the chemoreceptors. The introduction of time delay into models of the coupling path radically affects the limit-cycle behavior of the bidirectionally coupled system. Even the introduction of small time delays in the coupling paths can have marked effect on the limit-cycle stability. The dual-mode behavior of a reactively coupled system quickly changes to a single-mode condition as time delay occurs (Linkens, 1987).

COUPLING BETWEEN OSCILLATORS

Cells have physical contact with each other and the flow of information is achieved through the flow of chemical substances between them. An oscillatory behavior of one cell can therefore influence the behavior of a neighbor cell. Such coupling between oscillators will yield new results which are of interest in biology. It is possible to have sustained oscillations in the concentrations of chemical substances participating in various chemical reactions. Such oscillations could possibly serve as the biological clock. Both spatial and temporal time-dependent oscillations have been observed. There are many examples of periodically firing neurons in the nervous system of animals. Units of neuron fire burst periodically. One may visualize simple neuronal networks responsible for this behavior. Experience with most kinds of nonlinear oscillators indicates that a change in the parameters which causes an increase in amplitude results in a decrease in frequency.

A common feature of all models of neural networks is that the resulting periods of oscillations are determined to a large extent by the synaptic delays. Therefore, they cannot be much longer than the order of magnitude of these delays. It is unreasonable to expect that neuronal networks could produce oscillations of longer periods of oscillations. The nervous system may serve as a transducer rather than as a generator of oscillations. It is likely that the periodicity is generated at the subcellular level. There is no limit in the number of biological phenomena which are attributable to the properties of coupled oscillators.

Biological oscillators are complicated and a description of them using simple differential equations is probably not correct or adequate. All biological oscillators can be perturbed by suitable agents, like light, ions, or electromagnetic fields. The perturbation can show up as a change of oscillation with respect to stationary-state conditions. A phase change, a frequency change, or an amplitude change can occur as a result of such perturbations. A change in amplitude might be the result of an altered basic oscillator but could as well mean that a perturbation has occurred in the reaction sequence leading from the basic oscillator to the measurable rhythm. Persistent changes in the frequency or phase must be due to changes in the basic oscillator itself.

The analysis of rhythms with the phase-response curves and limit cycles has become fruitful (Pavlidis, 1973). The phase-response curve has been utilized to predict how the phase shifts due to repeated light pulses should look. Experimentally, it is observed that the eclosion rhythm locks to such pulses and adapts its 24-hour rhythm to the rhythm of 15-minute light pulses. This locking of an oscillator to external pulses is called the *entrainment*. Entrainment is only possible within a certain period interval around the natural period of oscillation.

The harmonic oscillator has a stable equilibrium point which is the origin. On the other hand, the Van der Pol oscillator equilibrium point is unstable. When x equals zero and W equals unity, in Equation 6.64, it approaches the expression

$$\{d^2x/dt^2\} - @\{dx/dt\} + x = 0 \tag{9.1}$$

The solution of Equation 9.1 has an increasing amplitude. This is an example of a situation that an equilibrium point, called the *point of singularity*, can be stable or unstable. Therefore, an important question to be answered is whether or not complicated biological oscillators can be perturbed in such a manner that the point of singularity is reached. In a controlled system with feedback, an oscillation can arise when the amplification is large enough and sufficient time delay is present.

PROPERTIES OF COUPLED OSCILLATORS

Consider two oscillators with characteristic frequencies, w_1 and w_2, which are weakly coupled and can be mutually entrained but with long transients. The behavior of the system is controlled by the sum of their outputs. Whenever the latter exceeds a threshold T, activity starts. Assuming that their outputs are sinusoidal,

$$y(t) = \sin[w_1 t] + \sin[w_2 t] \tag{9.2}$$

and that the difference $[w_1 - w_2]$ tends to zero, as time increases, a gross simplification can be obtained. Equation 9.2 can be written as

$$y(t) = 2\sin\{[w_1 + w_2]t/2\} \cdot \cos\{[w_1 - w_2]t/2\} \tag{9.3}$$

which is an oscillation with beats. A change in the period of oscillation as a function of time, as measured by the time intervals between successive events, can be demonstrated to occur. This is one possible explanation of the long and monotonic transients which can be described as a temporary loss of synchrony among the various units. This results in frequency beats modulating the free-run period.

If there are n cells in the population, they correspond to the existence of n harmonic oscillators of the kind described by Equation 6.64, with $A = 0$. All oscillators are coupled in such a manner that the kth oscillator satisfies the following differential equation:

$$\{d^2 x_k / dt^2\} + w_o^2 x_k = r w_o^2 \sum_{\substack{j=1 \\ j \neq k}}^{n} x_j \tag{9.3}$$

This is a linear differential equation and as such does not have much in common with biological oscillators. It is convenient to assume that all oscillators have the same angular frequency and that the coupling constant between any two oscillators in the structure is described by the same coupling factor r. In spite of these simplifications, the system of Equations 9.3 enables one to understand some salient points.

Addition of all Equations 9.3 yields

$$\{d^2 y / dt^2\} + w_o^2 [1 - (n-1) \cdot r] y = 0$$

$$y = \sum_{j}^{n} x_j \tag{9.4}$$

Therefore, y can oscillate with a frequency

$$\{w_o [1 - (n-1) \cdot r]^{(1/2)}\}. \tag{9.5}$$

This frequency is reduced with respect to the frequency of the individual oscillator when $r > 0$.

The population of the oscillators can oscillate also with other frequencies. Studying the difference $x_{k+1} - x_k$, for different values of k, yields

$$\{d^2[x_{k+1} - x_k]/dt^2\} + w_o^2(1 + r)[x_{k+1} - x_k] = 0 \qquad k = 1, 2, \ldots, (n - 1)$$
$$(9.6)$$

Therefore, under certain circumstances, the system can oscillate with a frequency $\{w_o(1 + r)^{(1/2)}\}$. If the number of oscillators is even, two subgroups of oscillators can arise with a phase difference of 180 degrees, indicating that the subgroups are oscillating out of rhythm. It should be evident that similar analysis can be extended to populations of nonlinear self-sustained oscillators with different frequencies and coupling factors.

Two different subgroups of oscillators can oscillate 180 degrees out of phase with each other. When all oscillators are synchronized, one event per cycle will be noticed. If the oscillators are divided into two subgroups, the number of events will be doubled. Such splitting of endogeneous rhythms has also been observed experimentally.

The splitting of a group of oscillators into subgroups can in principle show a hysteresis effect. When a coupling is changed until splitting occurs and then altered back to its original value, the splitting might still be present. A stronger coupling may be necessary in order to obtain the normal unsplit population again.

The relation in Equation 9.5 for the frequency of oscillations shows that when n is decreasing and r is greater than zero, the frequency should increase. The frequency of the periodic impulse activity of retina preparations of Aplysia californica also increases as parts of the retina are cut away. This influence of the population magnitude of the frequency is a further example of how useful the theory of coupled oscillators can be in the investigations of biological rhythms.

The frequency decrease realized in connection with the population of oscillators (Equation 9.3) enables one to explain the overall rhythms with long periods as a result of a coupling between individual oscillators with inherent short periods. Biochemical oscillations with short periods have been observed and investigated. The frequency reduction provides an attractive possibility to connect short-term biochemical oscillations with long-period rhythms. A population of coupled oscillators will be less sensitive to parameter changes which occur in the individual oscillator.

The oscillators can be coupled to each other in different ways. A perturbation of a population of interacting oscillators can result in a permanent change in the coupling and in the frequency. This could explain why frequency changes occur in biological systems both spontaneously and due to perturbations from the environment.

KINETIC PREREQUISITES FOR THE OCCURRENCE OF OSCILLATIONS

Feedback occurs when a process acts kinetically upon itself. Thus, it is basically a closed chain of action which causes the well-known effects of self-enhancement in the case of positive feedback and self-inhibition in the case of negative feedback. A general agreement has been reached concerning the origin of oscillations in chemical and biochemical systems as a result of nonlinearities in the underlying network of chemical reactions. In the phase-plane diagram, in which the concentrations of two oscillating reactants are plotted against each other, the path of the oscillating system may appear as a closed line, that is, a limit cycle. The limit cycle can display different shapes which may vary smoothly curved circles to sharp-edged figures, like triangles or rectangles.

The following characteristics have been listed by Frank (1979) for the occurrence of oscillations in physicochemical systems.

1. Sustained oscillations can occur only in thermodynamically open systems, far from equilibrium.
2. Oscillatory systems always consist of more than one degree of kinetic freedom; that is, the description of their temporal behavior requires a corresponding set of simultaneous differential equations.
3. There exist extremely nonlinear relationships between driving forces and driven fluxes or reactions.
4. Oscillatory systems always contain unstable states.
5. Oscillations are the result of mutual kinetic coupling between processes being otherwise independent of each other.
6. Oscillations of physicochemical systems are always accompanied by periodically occurring spatial propagation processes. They are, therefore, temporal and spatial phenomena at the same time.

The theoretical studies of process dynamics are commonly based on linear stability analysis and computer simulation. For nonlinear systems where the existence and stability of periodic solutions are of interest, these methods alone may lead to uncertain results. Thus, proven mathematical theorems are necessary and only the stable periodic solutions guarantee the possibility of experimental observations.

Excitations of metastable states are of primary importance for biological activities. They arise through nonlinear restoring forces and may have trivial or nontrivial consequences (Frohlich, 1988). In the case of a simple harmonic oscillator, with a restoring force proportional to the displacement x and the potential energy being proportional to the square of the displacement from the equilibrium position, one has

Force $f = -ax$

Potential energy $V = (a/2)x^2 > 0$ (9.7)

Proportionality constant $a > 0$

The oscillations are harmonic and all displacements are linear. Addition of a nonlinear term to the force, proportional to the cube of displacement, results in

$$V = (a/2)x^2 + (b/4)x^4$$
$$f = -ax - bx^3$$
(9.8)

If $b > 0$, the system will still oscillate, although not harmonically. The qualitative behavior of the oscillator is not changed, so long as b is positive. However, if b is negative, then a qualitative change occurs. The restoring force equals zero, in this case, both when $x = 0$, and when $x = x_o$,

$$x_o = \pm\{a/(\text{magnitude of } b)\}^{(1/2)}$$
(9.9)

The potential energy now rises, with an increasing magnitude of x below x_o. Above this range, the potential energy decreases. The potential energy thus has a maximum at $x = x_o$. At higher energies the system does not oscillate.

On the other hand, if b is positive and if a is negative, while the restoring force remains zero at $x = 0$, one has the result that the potential energy has a maximum at $x = 0$, and it has two minimums with negative values,

$$V = (1/4)\{a^2/b\}; \qquad b > 0$$
$$\text{at } x = \pm x_o$$
(9.10)

Thus, oscillations arise when $a > 0$, $b > 0$; metastable state when $a > 0$, $b < 0$; bifurcation and chaos occur when $a < 0$, $b > 0$.

Similar situations can arise also in the case of biological oscillators. The various known equations for oscillators and their properties have been listed by Kaiser as follows (Kaiser, 1988).

---

† Bifurcation theory describes the way qualitative changes in behavior occur as some control parameter is varied. Often the analysis of such changes can be done by considering a local normal form, as in the saddle-node and in period-doubling bifurcations. As the control parameter, m, passes through zero, a homoclinic orbit is formed, which is a trajectory biasymptotic to an unstable stationary point. The homoclinic bifurcation destroys a periodic orbit. Such bifurcations cannot be described algebraically by a simple linearization argument near the periodic orbit or stationary point alone. The existence of a homoclinic orbit is structurally unstable. A small perturbation of the homoclinic system no longer has such an orbit.

1. Unforced, undamped linear harmonic oscillator:

$$\{d^2x/dt^2\} + k^2x = 0 \qquad (9.11)$$

The solution is a constant sine (or cosine) wave, whose phase and amplitude are determined by the initial conditions. The period is determined by the magnitude of the force constant parameter k.

The second-order differential equation,

$$\{d^2V/dt^2\} + K\{dV/dt\} + KGV = 0 \qquad (9.12)$$

has a solution, which for certain values of the constants, K and G, is approximately a simple decaying exponential, for very small values for G, while for other values of the constants, the solution varies sinusoidally with time. Thus, it appears to be plausible that for the intermediate values of the constants, the solutions will be a compromise between these two extremes.

2. Undamped, unforced, nonlinear anharmonic oscillator:

$$\{d^2x/dt^2\} + [k + px^2]x = 0 \qquad (9.13)$$

Analytical solutions of Equation 9.13 exist. The period depends on the amplitude of the steady wavy motion. Beat modes, higher harmonics, and multistability can occur in this case.

3. Damped, unforced, linear harmonic oscillator:

$$\{d^2x/dt^2\} + b\{dx/dt\} + k^2x = 0 \qquad (9.14)$$

Equation 9.14 and its solution has been discussed earlier. The solution is an exponentially damped sine wave or a nonoscillatory exponential decay of the amplitude. The period depends on the friction coefficient b.

In the case of a damped, forced, linear harmonic oscillator, the analytical solution is a superposition of the solution of the unforced oscillator and a particular solution of the whole system, with its external frequency.

4. Damped, unforced, nonlinear anharmonic oscillator:

$$\{d^2x/dt^2\} + b\{dx/dt\} + [k + mx^2]x = 0 \qquad (9.15)$$

Equation 9.15 is known as the Duffing oscillator.

5. Van der Pol oscillator:

$$\{d^2x/dt^2\} + [b + cx^2]\{dx/dt\} + k^2x = 0 \qquad (9.16)$$

For the Van der Pol equation oscillator, no analytical solutions are known. Transients to an asymptotically stable equilibrium in the Duffing case and to an asymptotically stable limit cycle in the case of the Van der Pol equation can be computed. A necessary requirement for the existence of a limit cycle is a nonlinear damping term. Besides harmonics, subharmonics occur, as well as quasi-periodic and chaotic states occur. Both the forced Duffing oscillator and the forced Van der Pol oscillator are good examples of nonlinear excited systems.

The harmonically forced Duffing equation,

$$\{d^2x/dt^2\} + w^2x + Bx^3 = F\cos(qt) \tag{9.17}$$

possesses, for $B = 0$, the periodic solution,

$$x = \{F/[w^2 - q^2]\}\cos(qt) \tag{9.18}$$

of period $[2\pi/q]$, assuming that the magnitude of q equals the magnitude of w. The equation for the first variation is

$$\{d^2y/dt^2\} + \{w + 3Bx^2\}y = 0 \tag{9.19}$$

which for B equal to zero, admits only nontrivial periodic solutions of periods $\{2\pi/w\}$, $[4\pi/w]$ and so on. The Duffing equation possesses a unique family of periodic responses for small values of B, each with a period of harmonic input (forcing function), which converges to

$$x = \{F/[w^2 - q^2]\}\cos(qt)$$

as B tends to zero.

SELF-SUSTAINED OSCILLATIONS IN NONLINEAR SYSTEMS

The determination of the solution of the Van der Pol equation by the graphical method of isoclines enabled mathematicians to realize that the major stumbling block for the solution of the problem of nonlinear oscillations has been solved in principle. The simple concept of periodic motion prevalent in the theory of oscillations was replaced by the concept of the limit cycle. As a closed trajectory in the phase plane means evidently a periodic phenomenon, the discovery of limit cycles was fundamental for the new theory of self-excited oscillations.

The closed trajectories represent periodic phenomena, since the singular points are identified with positions of equilibria. The significance of the three principal singular points is simple: the node characterizes an aperiodically damped motion, the focus characterizes an oscillatory damped motion, and the saddle point denotes an essentially unstable motion occurring.

In some cases, there also occur semistable limit cycles, characterized by stability on one side and instability on the other side. Physically, only stable cycles are of interest. The unstable cycles play the role of separating the zones of attraction of the stable cycles in the case when there are several cycles. Everything which oscillates in a stationary state is necessarily of the limit-cycle kind. What is really difficult is to ascertain whether a given differential equation has a limit-cycle solution. This difficulty is due to the fact that the form of the differential equation does not convey any information regarding the existence of the limit cycles. However, the solution does yield this information.

SPORADIC NONPERIODIC OSCILLATIONS IN THE ACTIVITY OF THE ENDOCRINE SYSTEMS _____

The endocrine system forms a complex feedback mechanism with the central nervous system (CNS). The endocrine glands secrete under proper innervation and their secretions in turn affect the thresholds of the neurons in the different parts of the central nervous system. Because of the feedback nature of the whole system, the endocrine–CNS system, the cause of fluctuations in secretion should be considered from a global point of view.

Rashevsky (1972) has presented the results of his study of the activity of endocrine systems and the conditions under which the assembly of endocrine glands can undergo nonperiodic oscillations. The mathematical theory of interactions of n endocrine glands leads to the situation that under certain specified conditions, periodic nondamped oscillations occur. The variables one has to deal with in this case are

- The n average concentrations C_i ($i = 1, 2, \ldots, n$) of the ith hormone in the cell of the ith gland.
- The n average concentrations of the ith hormones in the intracellular fluids of the ith gland.
- The $\{n^2 - n\}$ of the kth hormone in the cell of the ith gland.
- The $\{n^2 - n\}$ average concentrations C_{ki} ($i, k = 1, 2, \ldots, n$).
- The $\{n^2 - n\}$ average concentrations C_k^i ($i = k$; $i, k = 1, 2, \ldots, n$) of the kth hormone in the extracellular fluid of the ith gland.
- The n average concentrations C_i^b of the ith hormone in the bloodstream.

One thus has $[2n^2 + n]$ concentration variables. Accordingly, one has $[2n^2 + n]$ simultaneous differential equations. Of these n equations are nonlinear, while the rest are linear.

The nonlinear equations are those that govern the rate of change of n variables C_{ki}. This is due to the appearance of the functions f_{ki}, which denote the effect of the kth hormone in the cell of the ith gland upon the rate of change of concentration of the ith hormone with time. Linearization of these n equations leads to $2n$ linear equations. Of these, n are valid for small values of C_{ki} below a threshold value, while the remaining n equations are valid for larger values.

The characteristic equation of the system of $\{2n^2 + n\}$ equations is utilized to find the conditions when the equation has at least one root with a positive real part, while the real parts of all other roots of the other characteristic equation are negative. This will result in the existence of negatively damped oscillations for small values of an auxiliary variable z and hence of C_i. The amplitudes of these oscillations will increase. For large values of C_i, the oscillations will be all positively damped, their amplitude decreasing. Thus, somewhere in between there will exist a point $C^\star$ around which sustained undamped oscillations will occur.

The condition that the real parts of all roots of the characteristic equation be negative is that a constant b_o be positive, and all the determinants be positive. Therefore, if a single determinant is negative, there will be at least one root of the characteristic equation with a positive real part. If there are many roots with positive real parts, then one can have sustained oscillations with different frequencies. The endocrine system can, in this manner, exhibit sporadic increases or decreases of activity which are not strictly periodic.

A MODEL FOR BIOLOGICAL OSCILLATORS

In biological terms, one may consider a biochemical oscillator where once the concentration of a substance exceeds a certain level, a chain of reactions independent of the oscillator is initiated and that time can be estimated by observing their results. One can then attempt to obtain an estimate of the structure of the system by perturbing it.

If one assumes that one has an oscillator with a single degree of freedom described by a pair of nonlinear differential equations,

$$\{dr/dt\} = f(r, s)$$
$$\{ds/dt\} = g(r, s) \tag{9.20}$$

where r and s are the two state variables and f and g are to be chosen so that the system of the preceding two equations exhibits a limit cycle.

In order that the system described by Equation 9.20 has a limit cycle, it is necessary that the characteristic equation of its linearized form has two roots of the same sign. If r_c and s_c are the coordinates of the critical point,

$$f(r_c, s_c) = 0 = g(r_c, s_c) \tag{9.21}$$

In the linearized form, one has

$$\{dr/dt\} = \{df/dr\}r + \{df/ds\}s$$
$$\{ds/dt\} = \{dg/dr\}r + \{dg/ds\}s \tag{9.22}$$

The two roots will have the same sign if

$$\{df/dr\}\{dg/ds\} - \{df/ds\}\{dg/dr\} > 0 \tag{9.23}$$

The inequality of Equation 9.23 is a necessary but not a sufficient condition for the existence of a limit cycle. If a critical point is asymptotically stable, then there must be a region R around it so that if the point (r_o, s_o) belongs to R, then the trajectory starting from it tends toward (r_c, s_c) for t tending to infinity. In terms of a model for the circadian oscillator, this implies that once the state of the system is in R, the oscillations would damp out and could be started again only by an external stimulus.

Such a system would be evidently undesirable as a biological regulator, unless the likelihood that its state was brought in R was very small. For the overall dynamic response of the system, a good first approximation could be achieved by ignoring R and assuming that the critical point is unstable; that is, the roots of the characteristic equation have nonnegative real parts. If they are zero, this would imply the existence of a center and the possibility of more than one periodic trajectory, depending on the initial conditions.

In view of the superposition properties and the structural instability of the previously mentioned system (very small changes in their parameters can result in drastic changes in the system behavior), the possibility mentioned earlier is unlikely. Therefore, the real parts should be positive, or

$$\{df/dr\} + \{dg/ds\} > 0$$

It is also preferable that the curves $f(r, s) = 0$, and $g(r, s) = 0$ do not have more than one intersection, since this would imply the existence of another critical point in the plane. The existence of another critical point would offer the possibility that large enough disturbances could move the system outside of the attraction region of the limit cycle and hence damp out the oscillation.

The previously mentioned conditions are fairly general, and they should be satisfied for any system which is a candidate for a model of the basic circadian oscillator. The salient feature of a system of coupled nonlinear oscillators is the multiplicity of stable solutions which predict that after a possibly minor disturbance, one can observe a drastic change in the macroscopic dynamical behavior of the system, for example, doubling of the frequency of oscillations. From a biological point of view, one should expect that any rhythm regulator at the cellular level should consist of a group of coupled oscillators.

A good example of how a basic mechanism is modified to produce spontaneous oscillator activity is the sino-atrial node of the heart. If the function of the sino-atrial node is considered from the starting point of a normal propagation of action potential along the axon, it is clear that following the application of a current spike, the threshold potential is achieved and an action potential is produced. In the case of the sino-atrial node, the fundamental difference between depolarization in these cells and the function of the nerve axon is shown to be the existence of a leaky membrane which allows the conduction of sodium ions into the cell, which in turn leads to spontaneous depolarization. Each sino-atrial cell can thus be considered as an autonomous biological oscillator, controlled by dynamics which can be closely approximated by a modified version of the Hodgkin–Huxley model.

Another important feature of the spontaneously oscillating physiological systems exhibited by the sino-atrial node is that it comprises a colony of many thousands of cells which depolarize synchronously. The process which produces this phenomenon is called *entrainment* and is an important property of spontaneously oscillating physiological systems which interact. In the sino-atrial node, the rate of depolarization is varied by sympathetic and parasympathetic activity affecting the cell membrane conductances. This is an example of how a biological oscillator can be influenced by external factors. The factors which influence a biological oscillator are often the outputs from another biological oscillator which is in turn oscillating spontaneously.

In many biological systems the oscillators are more or less similar to each other. They cannot be strictly identical, since if they are identical, no external noise can be present. Then infinitesimal mutual coupling would be sufficient to cause perfectly ordered behavior. The theoretical understanding of the origin of collective rhythmicity is best obtained by studying the onset, as a kind of phase transition or a bifurcation. The natural frequencies of oscillators in a biological system may be distributed over a certain range. Coupling among oscillators usually favors mutual synchronization. Physiological timing probably depends on many other factors than just the circadian clock.

The inevitability of phase singularities, even among mutually synchronizing cells, is particularly transparent in the timing of energy transfer in yeast cells. Biochemical clocks are evidently dependent on the biochemical reaction mechanisms. In the mathematical analysis of complex chemical reactions, one usually assumes that all concentrations are uniform, and that the rates of change of the concentrations of intermediates can be set equal to zero. This is the so-called *steady-state assumption,* which enables one to calculate their stationary-state concentrations. These intermediates are not truly constant concentrations in closed systems. However, they can be maintained constant in open systems by flow of materials.

In isolated closed systems, chemical reactions progress toward the final stable state of equilibrium. No other stable state is permissible and thus, if oscillations occur, they tend to die out. In open systems, on the other hand, stable conditions can exist so long as the flow of materials goes on. These stable states could be either stationary or oscillatory. Such stationary states or oscillatory states can be maintained so long as constant input and outflow are maintained. More than one stationary state is also possible.

The vast number of biochemical oscillations are the result of the occurrence of coupled, simultaneous enzyme-catalyzed reactions, occurring in open system and maintained by a continuous supply of energy and matter. Oscillations occurring in the synthesis of enzymes at the cellular level have also been reported to occur. These effects are attributed to the induction and repression mechanisms.

1. Biochemical reactions are organized in forms of enzymatic cycles such as the cycle energy metabolism and the biosynthetic pathways, all implying multiple types of positive and negative feedback interactions.

It is known that spontaneous oscillations appear as a result of feedback effects in electrical circuits. Similarly, the existence of feedback loops in chemical reactions can lead to oscillations. Feedback occurs when a process acts kinetically upon itself. Thus, it consists of a closed chain of action which causes the effects of self-enhancement in the case of positive feedback, and self-inhibition in the case of negative feedback. If the output acts upon the input of the same system, a feedback situation arises in which the effect is influencing its own cause. This kind of feedback has no effect on the properties of the transmission system, and it is designated as *nonsystemic feedback*. In chemical systems, nonsystemic feedback is brought forth by stoichiometric autocatalysis.

In physicochemical systems, most feedback mechanisms, however, act not upon the input but instead upon the transmission system itself. This kind of feedback is designated as *systemic feedback*. Systemic feedback is realized by reactions whose activation energy or rate constant depends on their own reaction products or reactants. Oscillatory systems contain at least two simultaneous processes. Feedback processes need time for their proceeding. The feedback delay which occurs is of essential importance for all temporal phenomena of feedback systems.

2. Many chemical transformations are controlled by allosteric enzymes responding to small changes in concentrations of substrate, products, and other controlling ligands in a cooperative manner by changing their conformation states.

3. The organization of the transport of chemical species in living systems implies not only free diffusion but to a large extent processes which occur upon coupling between transmembrane events and enzyme functions. The activity of membrane-bound enzymes is controlled not only by the components of the membrane itself, but in addition by the nature of membrane protein-directed diffusion processes.

TIME HIERARCHY IN OSCILLATING METABOLIC SYSTEMS

From enzymatic systems to ecosystems, one comes across many situations where the behavior of the system shows noised oscillations. Because of the intrinsic random character of the observed phenomenon, it has been suggested that deterministic models have less explanatory power than a stochastic model. The larger the variance of the noise, the more the behavior of the noised system is different from that of the deterministic system. One can obtain random oscillations from a weakly attractant stable focus.

Metabolic systems are composed of many biochemical reactions and transport processes with very different rates. Their dynamic interaction produces a temporal organization, whose salient feature is the existence of a distinct time hierarchy. Only a subset of the dynamic variables of the system moves with a velocity comparable to the characteristic time scale of the whole metabolic system, while others move very quickly and produce a dynamic rapid substructure (Dvorak and Kubinova, 1980). The description of this situation in mathematical form leads to differential equations with small parameters acting as multipliers of the derivatives of the fast components.

In the simplest case, one has two equations of the kind

$$\{dx/dt\} = f(x, y)$$
$$@\{dy/dt\} = g(x, y)$$

(9.24)

where $@$ is a small positive definite parameter. The initial conditions, $x(t = 0) = x_o$ and $y(t = 0) = y_o$, should be specified in order to solve this system of equations.

Since f and g are generally nonlinear functions, the solution of Equation 9.24 is a difficult task. When $@$ is sufficiently small, it is reasonable to approximate the solution by the solution of the *degenerate system,* (when $@ = 0$),

$$\{dx^\star/dt\} = f(x^\star, y^\star)$$
$$0 = g(x^\star, y^\star)$$

(9.25)

In biochemical systems, this is called the *quasi-steady-state approximation.* The steady states of the metabolic systems may be unstable. As a result, all variables may undergo continuous oscillations of more or less irregular, nonlinear character. Product activation of a key reaction in, the metabolic pathways can be considered an example.

Consider a reaction scheme in which the product activates the central reaction, which is the enzymic conversion of a substrate S to a product P. The substrate as well as the product are assumed supplied to the system at a constant rate. The product can flow freely out of the system, outflow rate being proportional to the concentration of P, while the substrate cannot. In addition, the product activates the rate of its own formation. The enzyme is also supplied to the system at a constant rate. The outflow of the enzyme from the system is inhibited by the product P. Consider that this outflow as a degradation of the free enzyme molecules, while those bound to the product are protected against degradation. The dynamics of such a system can be described by the following set of differential equations:

$$\{de/dt\} = 1 - e\{1 - [p^2/(p^2 + 1)]\}$$
$$@\{ds/dt\} = v_s - e\{[sp^2/(p^2 + 1)]\} \qquad (9.26)$$
$$@\{dp/dt\} = e\{[sp^2/(p^2 + 1)]\} - p + p_o$$

where the small letters are utilized to denote the concentrations of the corresponding species, identified by capital letters. v_s and p_o are normalized input flows of the substrate and the product, respectively. The input flow of the enzyme is normalized to unity. The coefficient @ is given by the ratio of characteristic times of the substrate and enzyme turnover.

The system of Equations 9.24 can be decomposed into two subsystems, a slow and a fast subsystem of the remaining two equations, when @ is very small. Linear stability analysis of the fast subsystem on the assumption that e is constant shows that there exists a large area of values of e and v_s where the steady state of the fast subsystem is unstable. Since the diverging trajectories cannot leave a certain bounded region, a limit-cycle behavior appears in the dynamics of the fast subsystem. Systems exhibiting such fast subsystem limit cycles, do not satisfy a theorem (known as Tikhonov's theorem), which justifies the quasi-steady-state approximation.

Consider the general systems whose trajectories of their fast subsystems are attracted by limit cycles for any fixed values of slow variables. These are typified by equations in the set of Equations 9.24 and 9.26. Suppose that for any fixed value of x the dynamic system generated by the fast equation

$$@\{dy/dt\} = g(x, y)$$

the trajectories approach a hyperbolic attractor. Dvorak and Kubinova (1980) discuss the derivation of a theorem which states that under appropriate conditions the solutions of the system of Equations 9.24, as specified in the foregoing, may be for @ tending to zero, arbitrarily closely approximated by the solution $x^\star(t)$ of the equation,

$$\{dx^\star/dt\} = \int_{A_x} f(x^\star, y)\, du_x(y)$$

with an initial condition $x^\star(o) = x_o$. The measures $u_x(y)$ are the Bowen–Ruelle measures on the attractors, A_x. The theorem of Bowen–Ruelle specifies the necessary conditions for the averages to be independent of the initial conditions of the trajectories. The reader is referred to the papers of Dvorak and Kubinova (1980) for additional details.

---

† The definition of an attractor is based on the intuitive property of invariance in two reciprocal operations consisting first of searching its basin and after taking the limit of this basin, when time tends to infinity.

GLYCOLYTIC OSCILLATIONS _____

Sustained oscillations have been observed for glycolysis. Evidently, the biochemical controls of glucose utilization are coupled intimately to the organism's need to derive energy from glucose oxidation. Theoretical studies attribute such oscillations to the enzyme phosphofructokinase, PFK, which is an allosteric enzyme and may be activated by the products ADP and FDP and inhibited by the substrate ATP (adenosine triphosphate).

In yeast and other cells that utilize oxygen, the rate at which sugar passes through the glycolytic pathway is controlled by a third enzyme, the phosphofructokinase, PFK, that attaches a second phosphate group to the fructose sugar molecule. When an adenosine phosphate pool of the cell, consisting of the mono-, di-, and triphosphate derivatives, PFK activity increases as much as a hundredfold. In the absence of those energy-poor phosphates, PFK activity shuts down. This shut down by oxygen is called the *Pasteur effect*.

In the Pasteur effect, the biochemical correlate of the organism's energy needed is reflected in the level of ATP or in the {ATP/AMP} ratio. Thus, the entry of glycosyl residues into the glycolytic pathway is determined to a large extent by the ATP generated from the metabolism of glucose residues which have traversed the glycolytic sequence and have entered the tricarboxylic acid cycle. Since hypoxia would diminish the intracellular concentration of ATP, the entry of new glucose molecules into the glycolytic sequence would be enhanced. The Pasteur effect is thus an expression of a servomechanistic biochemical control.

The role of PFK is emphasized by the fact that its substrate, fructose-6-phosphate, clearly exhibits the relatively large concentration amplitude of all oscillating glycolytic intermediates. The function of PFK is further supported by the phase-angle analysis of the amplitudes of concentration oscillations of all glycolytic intermediates as well as by phase-titration experiments. These identify the phase–shifting properties of those ligands which control the activity of PFK. Thus, this enzyme runs periodically through a state of high activity, alternating with a state of low activity. The periodic change is propagated along the glycolytic pathway through the adenine nucleotide system, which also affects the enzymes phosphoglycerate kinase and pyruvate kinase periodically. These two enzymes generate the product ATP, which feeds back to the enzyme PFK. PFK Not only operates in a feedback cycle via the adenine nucleotide system but also through an intrinsic feedback structure supplied by the oligomeric composition and nonlinear allosteric function. The regulatory mechanism of the enzyme PFK itself has autocatalytic features which are of importance for generating oscillatory activity.

When PFK is active, it is catalyzed by the triphosphate ATP and nibbles a phosphate group off, degrading it to a lower-energy species. This autocatalytic step is the root of the instability of the steady state. This instability evolves into a bounded oscillation. Based on this kind of a biochemical autocatalytic set of reaction mechanisms, the yeast cell behaves like a biochemical clock. It responds to chemical interventions that engage in its fundamental mechanism. Frequently, the only change that accompanies such perturbations is a phase shift itself. Glycolytic oscillations is a useful case to illustrate the boundaries of the dynamic domains of nonoscillatory as well as oscillatory states. Basically, its analysis shows that the dynamic state of the complex process can be reduced to the molecular properties of a single enzyme. Glycolytic oscillations have been long studied as an example of biochemical rhythms. Models based on the regulatory properties of the phosphofructokinase show that an instability-generating mechanism consisting of a single autocatalytic enzyme reaction operating far from equilibrium is sufficient to explain the periodic activity of glycolysis.

Many chemical transformations are controlled by allosteric enzymes responding to small changes in the substrate concentrations, products, and other controlling ligands in a cooperative manner, by changing their conformation states. Periodic phenomena may exist due to the hysteretic response of an enzyme conformation towards a change of controlling ligands. Furthermore, the kinetic delay of the production or consumption of the chemical species in a complex enzyme-reaction cycle can also result in oscillations. Models have been developed by Sel'kov (1968) and Higgins (1964) to represent observed oscillations in glycolysis. Sel'kov was able to demonstrate that the experimental data can be interpreted in terms of a mechanism involving an unstable transition point and a limit cycle thereafter. The allosteric character of PFK is taken into account through a phenomenological factor. Segel (1980) has presented the details of a model of Goldbeter et al. (1972, 1976) for the oscillatory PFK reaction, which takes into account the allosteric properties of the enzyme. The details of this analysis are lengthy.

Briefly, the model results in the formulation of the differential equations

$$\{dA/dt\} = S_1 - S_M F$$
$$\{dC/dt\} = k_s[LF - C] \tag{9.27}$$

where A and C are dimensionless concentrations. F is a known function of A, C, and other parameters of the system. The definitions of L, F, and k_s, which are lengthy, are not presented here. They are available in Segel (1980). The allosteric model is based on the transition mechanisms proposed by Monod et al. (1965) for multisubunit enzymes. The existence of cooperative interactions between the enzyme subunits and positive feedback are essential for a nonequilibrium instability in the PFK reaction. The various kinetic equations lead to the formulation of Equation 9.27. The dynamics of the system of Equations 9.27 in the phase plane $\{A, C\}$ is utilized to analyze the sustained oscillations and excitability of the reaction system. A necessary condition for the excitable and oscillatory behavior is that the nullcline $C = LF$ should be an S-shaped sigmoid in the phase plane.

If the perturbations in $[A]$ and $[C]$ are denoted by x and y, respectively, so that

$$[A] = A_o + x$$
$$[C] = C_o + y$$

(9.28)

one obtains by substitution,

$$\{dx/dt\} = -S_M\{\partial F/\partial A\}_o x - S\{\partial F/\partial C\}_o y$$
$$\{dy/dt\} = k_s L\{\partial F/\partial A\}_o x + k_s[L\{\partial F/\partial C\}_o - 1]y$$

(9.29)

where the subscript o denotes quantities of the steady state. The set of Equations 9.28 is linear and hence admits solutions of the kind

$$x = a \exp(wt)$$
$$y = b \exp(wt)$$

(9.30)

Substitutions of these in Equation 9.29 yields a homogeneous algebraic system of equations of the first degree for a and b. In order to have a nontrivial solution, the determinant should be zero. This is given by the characteristic equation for w,

$$w^2 + w\{S_M(\partial F/\partial A)_o + k_s - k_s L(\partial F/\partial C)_o + S_M(\partial V/\partial A)_o = 0 \qquad (9.31)$$

The determination of the stability property of the steady state thus reduces to the analysis of the real part of the solutions of Equation 9.31, that is, whether w has a positive or negative real part.

This periodic behavior or anaerobic sugar metabolism can be obtained by solving the known equations of regulation within this cycle of interacting enzyme reactions. Glycolysis is the main energy-producing degradative pathway in cells that lack oxygen. In cells that utilize oxygen, the Krebs cycle takes place.

It may be stated that the glycolysis cycles, NADH (nicotinamide adenine dinucleotide) produces some ATP, while the Krebs cycle produces more ATP utilizing the NADH available. Thus, in the presence of excess oxygen, concentrations of ATP are high while NADH concentrations are kept low, and there is no oscillation. With no oxygen or low amounts of oxygen, glycolysis dominates and NADH concentrations remain high, keeping the concentrations of ATP lower.

Circadian rhythms show many characteristics which are typical of oscillators of the Van der Pol kind. The circadian rhythms are self-sustained oscillations and endogenous. However, they can be influenced by light–

dark conditions of the environment. In order to simulate the behavior of circadian rhythms, the following differential equation, based on the Van der Pol equation, has been suggested.

$$\{d^2y/dt^2\} + f_1(y)\{dy/dt\} + f_2(y)y = \{d^2x/dt^2\} + \{dx/dt\} + x \qquad (9.32)$$

where y corresponds to the oscillating biological variable. The functions f_1 and f_2 are functions of y which are utilized to account for the experimentally observed results about the rhythms. The right-hand side of Equation 9.32 describes how light intensity x excites the oscillator. Several properties of circadian rhythms can be simulated on the basis of this equation.

Circadian clocks behave like mechanical clocks in three respects: (1) their period returns to normal promptly after a stimulus; (2) by a suitably chosen stimulus they can be forced away from their regular behavior to a new phase; and (3) the amount of departure depends smoothly on the timing of the stimulus. However, resetting the curve need not be a mere distortion of the old phase. Mathematical models that exhibit such resetting types can be constructed. Models describing oscillatory phenomenon have been studied by many authors, mainly by computer calculations. All these models are given by a certain type of two first-order differential equations of the kind (Plesser et al., 1979):

$$\{dS/dt\} = V_{in} - V(S, P)$$
$$\{dP/dt\} = V(S, P) - V_{out} \qquad (9.33)$$

where the velocities V are functions of the concentrations of the substrate S and the product P. The input rate of the substrate and the output rate for the product are denoted respectively by V_{in} and V_{out}.

The introduction of dimensionless parameters s and p for the concentrations of S and P and a normalized time constant $t^{\star}$ enables one to rewrite Equations 9.33 as

$$\{ds/dt\} = q - v(s, p) = \dot{s}$$
$$\{dp/dt\} = @[v(s, p) - r(p)] = \dot{p} \qquad (9.34)$$

where @ is a scaling parameter, which is composed of some reference constants of the normalization procedure. q denotes the input rate.

Most biological oscillators are of the kind called *relaxation oscillators,* which has been discussed before. The Van der Pol oscillator provides an example of a nonlinear physical oscillator and has been the base for several models of biological rhythms. In order to simulate the behavior of biological rhythms, the following differential equation has been proposed.

$$\{d^2y/dt^2\} + F(y)\{dy/dt\} + G(y)y = \{d^2x/dt^2\} + \{dx/dt\} + x \qquad (9.35)$$

where y corresponds to an oscillating biological variable. F and G are functions of y, which are used to account for the experimentally observed results regarding rhythms. The right-hand side of Equation 9.35 indicates

the manner in which the system is excited by external perturbations such as
light intensity. Several properties of circadian rhythms can be simulated
based on this kind of equation. All biological oscillators can be perturbed by
suitable agents like electromagnetic radiation, ions, and so on. The perturbation can manifest as a change in phase, amplitude, or frequency. Persistent
changes in frequency or phase should be due to changes in the basic oscillator
of biological systems.

While a harmonic oscillator has a stable equilibrium point, the
equilibrium point of the Van der Pol equation is unstable. This is due to the
fact that as x approaches zero, the equation

$$\{d^2x/dt^2\} - @\{dx/dt\} + x = 0 \tag{9.36}$$

has the solution which has an increasing amplitude. Thus, the point of
singularity, the equilibrium point, can be stable or unstable. The limit cycle
of a simple oscillatory system contains at least one singular point. A nearly
complete analysis of the Van der Pol equation is presented by Struble
(1962). This essentially completes our brief overview of the rhythmic
phenomena in biology. In the appendices of this chapter, we present a
discussion of the autocorrelation functions and chaotic phenomena, which
are somewhat related to biological oscillations.

Appendix 9A: Autocorrelation function

The use of autocorrelation functions (Basar, 1976) and power spectral density
functions find applications in the study of biological oscillators. The
autorocorrelation function for random data describes the general dependence
of the values of the data at one time on the values at another time. An
estimate of the autocorrelation functions between the values $x(t)$ at time t and
$t + t^\star$ may be obtained by taking the product of the two values and
averaging over the period of observation, T. The resulting average product
will approach an exact autocorrelation function as T approaches
infinity. Mathematically, the autocorrelation function is defined as

$$R(t^\star) = \lim_{T \to \infty} (1/T) \int_o^T x(t) \cdot x(t + t^\star)\, dt \tag{9.A1}$$

The autocorrelation function is estimated by the following operations:
(1) the signal is delayed by a time displacement equal to the lag time $t^\star$; (2)

the signal value at any instant is multiplied by the value which has been recorded $t^\star$ seconds before, and (3) the instantaneous product value is averaged over the sampling time T. By moving the lag time $t^\star$ over the sampling time T, a plot of the autocorrelation function $R(t^\star)$ versus lag time is obtained.

The important application for the evaluation of the autocorrelation function of biological data is to determine the influence of the values at one time over values at a future time. The autocorrelation function can be viewed as an average measure of the relation of the values of random processes at one instant of time to the value at another instant of time $t^\star$ seconds later. A sine wave will have an autocorrelation function which persists over all time displacements. In this manner, an autocorrelation function provides a powerful tool for detecting deterministic data which might be masked in a random background.

The autocorrelation function can be considered as a measure of the frequency content of the sample function. The measure of the frequency content can be displayed more explicitly by forming the Fourier transform of the autocorrelation function. For stationary processes, the two functions, the power spectrum and the autocorrelation function, are related by a Fourier transform (Wiener–Khinchin relation). Any study of the oscillatory behavior of biological systems should include use of the previously mentioned functions. The autocorrelation function yields a direct measure for the periodicities of the studied signals.

Appendix 9B: Chaos in biological systems

The concept of chaos in biology and its implications for living systems are being recognized recently. It seems plausible that certain neurological disorders as well as cardiac arrythmia can originate from chaotic dynamics. It has been only recently recognized that chaotic behavior is a well-defined property in the dynamics of nonlinear systems (Chandra, 1984).

Every beat of the human heart is controlled by the sino-atrial pacemaker of the sino-atrial node located in the right atrium. The pacemaker imposes its own rhythms on the rest of the heart, generating electrical impulses to the first atria, which contracts to fill the ventricles with blood, and then to ventricles, which contract and pump the blood through the body. The initiation and maintenance of the heartbeat by its natural pacemaker is a kind of democratic process in which thousands of pacemaker cells, all capable of generating electrical signals, communicate electrically, reach a consensus, and discharge synchronously to initiate each heartbeat.

A cardiac arrythmia is an abnormality of the rate of rhythm of the heart. "Extra beats" are examples of cardiac excitation occurring prematurely. Sometimes, they occur as *ventricular tachycardia* or arrythmia in which the heart rate is above the normal range of 60 to 100 beats per minute. At other times, *ventricular fibrillation* or an arrythmia in which the usual coordinated contraction of the heart muscle is replaced by an extremely fast, asynchronous, and unpredictable activity. Insight into the mechanisms of these and other life–threatening arrythmias require a detailed knowledge of the electrical activity of the heart tissue.

The mechanisms of lethal rhythm disturbances exhibits irregular periodicities as well as an exquisite sensitivity to the initial conditions. These could be studied from the theory of nonlinear dynamical systems. Sudden death occurring after myocardial infarction, due to localized tissue damage as well as abnormal impulse formation may also be understood from such studies. Traditionally, the term *chaotic* has been used by cardiologists to describe irregular cardiac activity in which the excitation of the individual cells occurs in a seemingly random and unpredictable fashion.

Chaotic behavior is inherent in the solutions of many deterministic equations of dynamical systems. The inherent behavior arises when some solutions have an extremely sensitive dependence on their initial conditions.

In a dissipative system, all orbits may eventually be attracted to a stationary point, for example, the point of lowest energy, when a dissipative perturbation is added to a Hamiltonian system. A second possibility also exists of a stationary state, limit cycle, or a simple periodic attractor. In the case of the Van der Pol equation, as one changes the value of a certain parameter, u, which in several systems plays the role similar to that of the Reynold's number in fluid mechanics, one finds that a one-loop periodic attractor of period T changes at some value u_1 into a double-loop attractor of period $2T$. At some u_2, it acquires four loops and doubles its period again (Helleman, 1980). The u_k's at which these *period–doubling bifurcations* take place often converge to some finite critical value, u_∞, producing more and more complicated attractors in the process. Beyond u_∞, the object is no longer an attractor, but other nonperiodic attractors can arise with complicated shapes in the phase space. The motion along these attractors can be chaotic, ergodic, and even mixed.

Models of chaotic systems are generally nonlinear equations that cannot be solved analytically and they tend to lose all their nontrivial properties at linearization. As computer solutions are easily obtained, one is tempted to skip the mathematics completely. In chaos, a stochastic system originates from a deterministic system. A *deterministic system* is one whose past and future are both unique functions of the present. The future is a unique function of the present, but there are two different pasts of equal probability. One of the intriguing problems in nonlinear dynamics is that of understanding how simple deterministic equations can yield apparently random solutions.

The equations of motion of chaos are entirely deterministic, and there are no random inputs. Chaotic systems exhibit sustained motion. They do not settle down to either equilibrium or simple cycles. One observes solutions which never repeat and are often highly irregular. They exhibit sensitivity to initial conditions. Nearby trajectories on average separate exponentially and small differences in initial conditions are amplified. Since one can never specify a system's state with great precision, long-term forecasting becomes impossible. Often a succession of dynamical states is observed as one varies a parameter. These transitions called bifurcations can lead to chaos. Chaotic behavior arises out of a cascade of period-doubling bifurcations (Degn et al., 1986). Chaotic motion often indicates a noninteger or fractal dimension.

Such a chaotic dynamic system has presumably many solutions which display highly aperiodic or erratic time dependence. Consider the equation for a forced pendulum:

$$\{d^2X/dt^2\} + \sin X = @ \cos wt \qquad (9B.1)$$

where w is a constant angular forcing frequency, and $@$ is a small parameter. For $@$ very small but not equal to zero, Equation 9B.1 possesses no analytical integrals of the motion (Marsden, 1984). It possesses transversal intersecting stable and unstable manifolds. This type of dynamical behavior has several consequences leading one to use the term *chaotic*. Equation 9B.1 has infinitely many periodic solutions of an arbitrarily high period.

A number of problems arise in connection with chaotic behavior.

1. Is an apparently periodic state found in numerical calculations really chaotic, or is it quasi-periodic, or periodic with an extremely long period?
2. How can one separate noise and uniform randomness from deterministic irregularities and structured chaos?
3. What is the origin of deterministic chaos and how can one measure the strength of chaos?

A nonlinear oscillator that is subjected to an external periodic forcing can exhibit stable and unstable periodic and almost periodic responses, coexisting subharmonic states, and the onset of a completely irregular motion. This complex dynamical behavior seems to be an aspect of biological regulation. It is found in very simple nonlinear systems. Waves that behave like particles, called *solitons*, transitions within regular motions and from regular motions to irregular ones, the formation of spatial or temporal spatial structures from an isotropic medium are examples of such nonlinear phenomena. Within *chaotic states*, a large number of periodic states exist. Small parameter changes can lead to bifurcations from regular to irregular motions and vice versa.

Some criteria have been presented for a chaotic motion. The occurrence of chaos corresponds to a qualitative change in the behavior. The Kolmogorov entropy is considered a fundamental measure for a chaotic motion. It gives the average rate at which information about the state of the dynamical system is lost with time. For a regular motion, the Kolmogorov entropy K becomes zero, and for random systems it is infinite. Deterministic chaos exhibits small, positive K values. K values give some insight into the predictability of the system. The Liapunov exponent measures the average divergence of nearby trajectories. The fractal dimension is also closely related to the Liapunov exponent and the Kolmogorov entropy. The Liapunov exponent changes with internal or external parameters and represents a generalization of linear stability analysis. Systems with at least one positive Liapunov exponent are chaotic. These are the relevant measures for an identification of the deterministic chaotic states, particularly for an analysis of the experimental signals.

Systems of differential equations of the form

$$\{dX/dt\} = F(X) \tag{9B.2}$$

describe an oscillator which has a stable limit cycle. The nonlinear function F describes the development of the system with time. The system described by Equation 9.36 may consist of m ordinary differential equations of the first order. The m-dimensional first-order differential equations may be transformed into one ordinary differential equation of order m. A collection of such oscillators, weakly coupled by nearest neighbor coupling, can *phase lock*. For identical oscillators, this means that the homogeneous solution is asymptotically stable. However, if there is a variation in frequencies of these oscillators, then the coupled system must balance two conflicting tendencies:

1. The tendency to coherence encouraged by coupling.
2. The tendency to dispersion produced by the variation in the underlying frequencies.

It has been demonstrated (Koppell, 1984) that symmetry properties help determine the size of the gradient in the frequency that can be sustained without loss of oscillator coherence. When there is phase locking, such properties affect the frequency at which the ensemble runs. For frequency gradients which are too large to allow phase locking, local coherence remains after global coherence is lost.

A fundamental question is what mechanism in these two categories of systems leads to chaotic oscillations? Are the mechanisms similar or are they distinct? The chaotic oscillations arise as a result of instability of the system with respect to finite large disturbances in one case, and arises from instability with respect to infinitesimal disturbances in the other case. Self-excited chaotic oscillations appear to arise from a somewhat different mechanism than that associated with forced oscillations. In the case of the Lorentz equations, such chaotic oscillations appear to be the result of all static and dynamic equilibrium solutions becoming unstable with respect to infinitesimal disturbances.†

Nonlinear systems exhibit an enormous variability of behavior. If one changes at least one internal or external parameter, the system undergoes continuous or discontinuous changes from one attractor to another at some

---

† The Lorentz equations are:

$$\{dx/dt\} = s(y - x)$$
$$\{dy/dt\} = -xz + rx - y$$
$$\{dz/dt\} = xy - bz$$

where s, r, and b are parameters of the system. x, y, and z are functions of time. Much of our understanding of these equations and their solutions is summarized in Sparrow (1982). For $r < 1$, it is easily shown that only one single static equilibrium solution exists. For $r > 1$, three static equilibrium solutions exist. The equilibrium described by

$$x_o = y_o = z_o = 0$$

is stable with respect to infinitesimal disturbances for $0 < r < 1$ and unstable for $r > 1$. For additional details see Dowell and Pierre (1984).

Kolmogorov entropy, K, arises in connection with probability and random variables. It is a fundamental measure for a chaotic motion and gives some insight into the predictability of the system. Kolmogorov entropy gives the average rate at which information about the state of a dynamical sytem is lost in time. For regular motion, K becomes zero, and for random motion K is infinite. Deterministic chaos exhibits a small positve K value (Grassberger and Procaccia, 1983).

The Liapunov exponent changes with the internal and external parameters. It represents a generalization of linear stability analysis. Systems with at least one Liapunov exponent greater than zero are chaotic. The Liapunov exponent L, which measures the average divergence of nearby trajectories, is computed as follows for the Ricker equation and logistic equation:

$$X_{t+1} = X_t \exp[r\{1 - X_t\}]$$
$$X_{t+1} = rX_t\{1 - X_t\}$$

where r_t may assume different values in a periodic or random sequence.

$$L = (1/N) \sum_{t=1}^{N} \log_2[dX_{t+1}/dX_t]$$

The Liapunov exponent is an indicator of order when $L < 0$, or chaos when $L > 0$. The value of L thus can enable one to forecast the system behavior. For $L < 0$, uncertainties in the initial conditions are damped out. For $L > 0$, any uncertainty in the initial conditions will grow exponentially, thus making long-term predictions impossible.

critical values. (Stable limit cycles act as *periodic attractors*. If the trajectory never closes but the motion is restricted to a finite region, then it is called a *strange attractor*. Strange attractors represent a mapping of nonperiodic self-oscillations in the phase space of a dynamical system.) The determining parameter is called the *bifurcation parameter*.

One has three kinds of bifurcations: (1) Hopf bifurcation, (2) saddle-node or tangent bifurcations, which are transitions from one limit cycle to a new one, and (3) period-doubling bifurcations, when a limit cycle or period T bifurcates into a cycle of period $2T$ (Kaiser, 1988).

REFERENCES

BASAR, E., *Biophysical and Physiological Systems Analysis*. Reading, Ma: Addison-Wesley Publishing Co., 1976.

CHANDRA, E.J., *Chaos in Nonlinear Dynamical Systems*. Philadelphia: SIAM, 1984.

DEGN, H., A.V. HOLDEN, and L.F. OLSEN, *Chaos in Biological Systems*, NATO ASI series. New York: Plenum Press, 1986.

DOWELL, E.H., and C. PIERRE, in *Chaos in Nonlinear Dynamical Systems*, ed. J. Chandra, Philadelphia: SIAM 1984, p. 176.

DVORAK, I., and L. KUBINOVA, In *Mathematics of Biology and Medicine*. p. 300, eds. Capasso, Grosso, and Paveri-Fontana. Berlin: Springer-Verlag, 1980.

FRANK, U.F., in *Kinetics of Physicochemical Oscillations*, Plenary Lectures of a Disscussion Meeting, held by Deutche Bunsengesellschaft fur Physikalische Chemie, 1979, p. 675.

FROHLICH, H., *Int. J. Quantum Chemistry*, vol. 2, (1968), p. 641.

FROHLICH, H., *Biological Coherence and Response to External Stimuli*. Berlin: Springer-Verlag, 1988.

GOLDBETER, A., and R. LEFEVER, *Biophys. J.*, vol. 12, (1972), p. 1302.

GOLDBETER, A., and G. NICOLIS, *Prog. Theor. Biol.*, vol 4, (1976), p. 65.

GRASSBERGER, P. and I. PROCACCIA, I., *Phys. Rev.*, A: 28A: (1983), p. 2591.

HELLEMAN, R.H.G., *Nonlinear Dynamics*, Annals of New York Academy of Sciences, vol. 357. New York: The New York Academy of Sciences, 1980.

HIGGINS, J., *Proc. Nat. Acad. Sci., U.S.A.*, vol. 51, (1964), p. 989.

JALIFE, J., in *Mathematical Approaches to Cardiac Arrhythmias*, Annals of New York Academy of Sciences, vol. 591. New York: The New York Academy of Sciences, 1990.

KAISER, F., in *Biological Coherence and Response to External Stimuli*, p. 25, ed. H. Frohlich. Berlin: Springer-Verlag, 1988.

KOPPEL, N., in *Chaos in Nonlinear Dynamical Systems*, p. 86, ed. F.J. Chandra. Philadelphia: SIAM, 1984.

LINKENS, D.A., in *Control Aspects of Biomedical Engineering*, ed. M. Nalecz. Oxford: Pergamon, 1987.

MARSDEN, J.E., in *Chaos in Nonlinear Systems*, p. 19, ed. F.J. Chandra. Philadelphia: SIAM, 1984.

MONOD, J., J. WYMAN, and J.P. CHANGEUX, *J. Mol. Biology*, vol. 12, (1965), p. 88.

PAVILIDIS, Th., *Biological Oscillators: Their Mathematical Analysis*. New York: Academic Press, 1973.

PLESSER, Th., D. ERLE and K.H. MAYER, In *Kinetics of Physicochemical Oscillations*, Discussion meeting of Deutsche Bunsengesellschaft fur Physikalische Chemie, vol. III, p. 360. Aachen, 1979.

RASHEVSKY, N., *Bull. Math. Biology*, vol. 34, (1978), p. 65.

SEGEL, L.A., *Mathematical Models in Molecular and Cellular Biology*. London: Cambridge University Press, 1980.

SEL'KOV, E.E., *Eur. J. Biochem.*, vol. 4, (1968), p. 79; see also *Mol. Biol.*, vol. 2, (1968), p. 252.

SPARROW, C., *The Lorentz Equations: Bifurcations, Chaos, and Strange Attractors*, Applied Mathematical Sciences, vol. 41. New York: Springer-Verlag, 1987.

STRUBLE, R.A., *Nonlinear Differential Equations*. New York: McGraw-Hill Book Company, 1962.

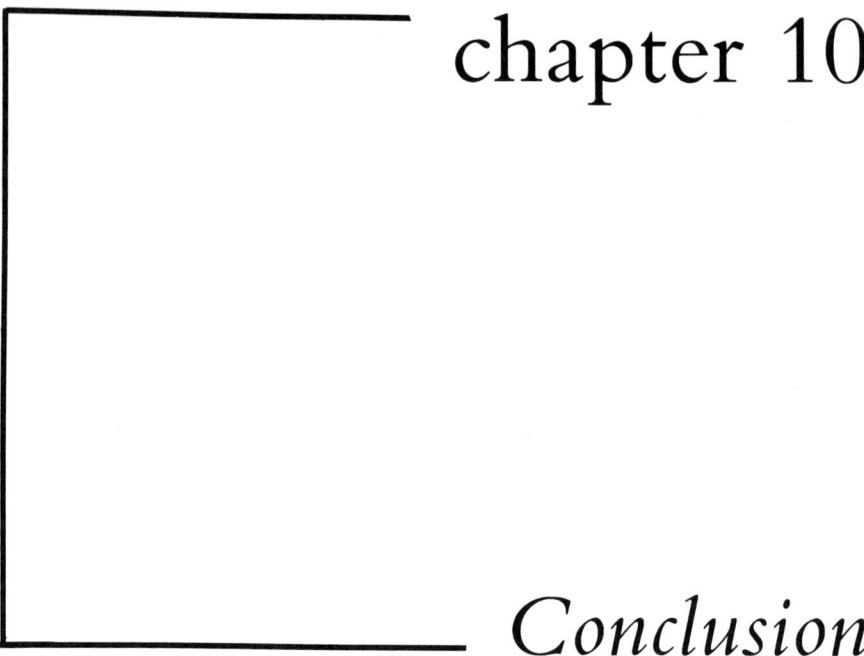

chapter 10

Conclusion

INTRODUCTORY REMARKS

Regulation is a fundamental property of living systems and is related to adaptation of the system to the changing environment. An important feature of the response of biological systems to an external stress when they are under homeostatic control is that these systems which are inherently nonlinear exhibit a discontinuity in their response to stress. A null effect at a strong stress does not imply an even smaller effect at a weaker stress. As commonly occurs in living systems under efficient homeostatic control, often larger biphasic effects are obtained when stress is applied. In this book we present an overview of the problems of understanding homeostasis from the point of view of a physical chemist.

In a living system, there are many kinds of microscopic components, such that one cannot pretend to develop a dynamical description of the system based on the properties of each component. Thus, theoretical physics has had a more difficult time in the understanding of living systems than with much simpler inanimate systems of physics. The generality of feedback phenomena giving rise to instabilities leads one to conclude that self-organization is an exception and not the rule. It appears to be impossible to describe a transition process for some physiological parameters without taking into account its interaction with other parameters of another functional system in terms of which this parameter is regulated.

Ideally, a physical theory should consist of a mathematical structure, together with certain rules of interpretation, encompassing the significant mathematical and physical aspects. In many physical problems, it is sometimes important not to know the specific value of the solution for some concrete value of the argument, but the kind of behavior for changes in the argument, and in particular for the boundless increase in the value of the argument. It is sometimes important to know whether the solutions which satisfy given values of the initial conditions are periodic or whether they approach some known function asymptotically.

Physics has elucidated many problems about the structure of complex systems by relating apparent macroscopic features of the system to the collective properties of microscopic components. Many chemical reactions and biochemical processes involve the transfer of electrons from one part of the system to another. In chemical processes involving only small molecules, the distance through which the electron is transferred is often quite small of the order of the linear dimension of the molecules involved. The passage of electron from one stable state to another stable state in this situation is explainable in terms of the *quantum mechanical tunneling across the barrier*. The probability of tunneling decreases exponentially with increases in distance over which the electron needs to be transported. In the analysis of the transport of electrons in biological systems, one has to consider the transport of electrons over a distance of about 100 angstroms or more. In this situation, the tunneling probability is almost zero. One has to invoke concepts from solid–state physics, like band theory.

In isolated polypeptide units, there are three distinct levels of pi–electrons. When N such polypeptide units are connected to form a linear protein, these three levels form three energy bands each containing N sublevels. The presence of a large energy gap between the valence and conduction bands results in such a system being an insulator, or a dielectric. In order that an electron is excited to the conduction band and then gets transported, energy of the order of four electron volts should be supplied, which is plausible only at high temperature, or by incidence of electromagnetic radiation. When such possibilities do not exist, one resorts to explanations like virtual transition. This means that the donor molecule polarizes the protein molecule, which in turn polarizes the acceptor molecule. Thus, the donor and acceptor molecules interact through the polarizability of the protein molecule. Hence, one can cause electron transport via the donor–polarizable molecule acceptor. Certain quantum mechanical considerations relevant to high–temperature superconductivity as well as Anderson Hamiltonian and soliton concepts allied to this problem are not presented here.

The generally accepted method of the storage of energy in biological systems is the formation of chemical compounds with energy-rich bonds, which are broken later when energy is needed for some other reaction. The well-known example is the ATP-ADP conversion. Two other physical methods of storing energy also exist. One is the formation of a perpetual current, as in superconductors, where the kinetic energy of electrons is maintained. The second possibility is the laser, where the energy is pumped and stored until population-level inversion takes place, when energy is released in a coherent fashion. The population-inverted state is metastable and higher in energy than the absolute-ground state.

It is possible to store energy in such metastable states far away from absolute-ground state. According to Frohlich (1988), this is possible in situations like cell membranes, where the high polarizability makes excitations of collective modes possible. The bilayer structure of membrane lipids gives rise to the possibility of collective modes which are found in smectic liquid crystals. Flexo-elastic effects are produced; this can affect the thickness of the membrane, which becomes thinner when polarized. (Polarization produces surface charges and elastic deformation, with the constraint that volume per unit mass remains constant. Conversely, elastic deformation produces polarization and surface charges.) This variation in membrane thickness while a polarization wave passes along the membrane must affect the transport of ions and species across the membrane. The postulation of these concepts and problems in quantum mechanical language is elegant. However, presentation of these concepts in quantum mechanical language in this book involves elaborate introduction of various principles and notations, which makes the presentation tedious and clumsy. Therefore, we present a much simplified classical picture in the following.

Frohlich (1988) suggested that analogous to the laser, a possible storage of energy in a coherent fashion exists in states far from the thermodynamic equilibrium state. The relevant quantum mechanical formulation can be stated classically as a case of a damped nonlinear oscillator problem:

$$m\{d^2q/dt^2\} + C\{dq/dt\} + Aq + Bq^3 = 0 \tag{10.1}$$

where m is the mass, C is a damping term proportional to velocity, A is a linear restoring force constant, and B is a nonlinear restoring force. q is the position variable of particle with mass m.

When the damping is heavy, that is, when C is very large, one has

$$C\{dq/dt\} = -Aq - Bq^3 = K(q)$$
$$K = -\{\partial V/\partial q\} \tag{10.2}$$

K is the restoring force derivable from a potential V. Then

$$V(q) = (1/2)Aq^2 + (1/4)Bq^4 \tag{10.3}$$

It should be noted that only even powers in q occur. Thus, V is symmetric.

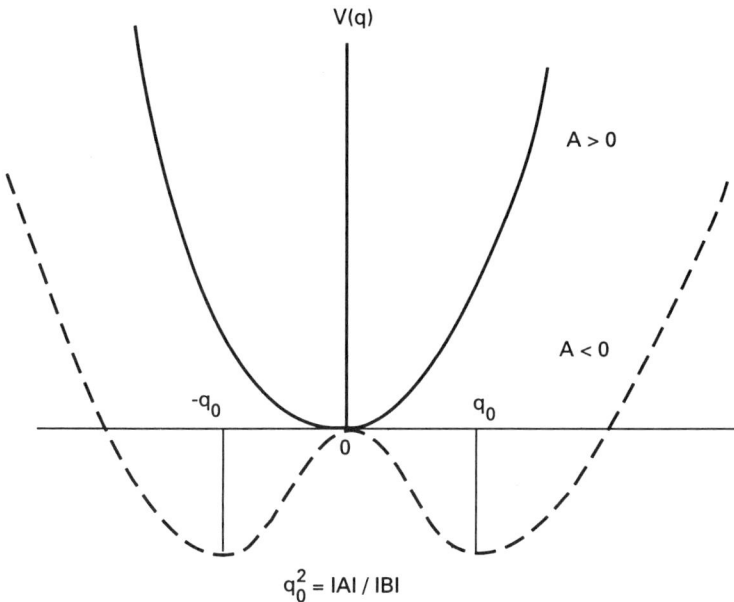

Figure 10.1 Potential profile as given by Equation 10.3, when B is positive definite and real. The real nature of B and its positive definiteness alone leads to stable oscillations.

If one assumes that B is real and is positive definite, the preceding equations result in stable oscillations. The nonlinear term provides the feedback necessary for stable oscillations. When A is positive definite, one has that the potential has one minimum at $q = 0$, which is a stable equilibrium. If A is negative, then there exists two minima, since the situation at $q = 0$ becomes unstable, and new minima occur at $q = \pm q_o$. The potential profile is presented in Figure 10.1.

As A tends to zero, the motion becomes sluggish as the restoring force is reduced. This is called *critical slowing* or *mode softening* and signals an imminent phase transition. When A changes sign from positive to negative values, the minima at $q = 0$ become unstable and new minima at $q = q_o$ evolve. The system breaks its symmetry spontaneously and attains a new ground state, whose symmetry is different from the earlier ground state.

If the nonlinear term is negligible and the damping term is still heavy, one has

$$\{dq/dt\} = -Aq$$

$$q(t) = q(o)e^{-At} \tag{10.4}$$

$$q(t) = q(o)$$

as A tends to zero. One may further make use of the linear response theory to compute various possible modes analogous to laser action, namely, from random inputs (pumping) to coherent output radiation.

Living systems are organized by the regulation of their internal degrees of freedom, such as concentration levels and fluxes of chemical species, as well as electrical effects. Control is exercised through electrically sensitive structures, such as cell membranes. Homeostasis is maintained over a wide range of perturbations. Many biochemical networks of chains of reactions which are parametrically unstable, are organized. Periodic processes are one of the most important functional characteristics of the biological systems. Both observations and theory suggest that oscillatory behavior is a general and perhaps a necessary characteristic of the homeostatic mechanisms of living systems. Usually, there is a logarithmic response for a stimulus.

REFLECTIONS OF NONEQUILIBRIUM THERMODYNAMICS

The phenomenological description of systems with the aid of nonequilibrium thermodynamics is usually made with the help of a set of macroscopic quantities, so chosen that their future values are unambiguously determined by their values at the present time. One assumes from the outset that a phenomenological description is possible, and a characteristic set of macro-quantities exists. For example, concentrations and chemical potentials are considered as fundamental quantities. The applicability of the phenomenological description as well as the existence of local equilibrium conditions to situations far from equilibrium is assumed to be valid. The thermodynamics of biological processes ought to describe all fundamental biological phenomena from a single standpoint, so that it can form a reliable basis for theoretical biology. The main difficulties in the formulation of a system of thermodynamics for biological processes are concerned with the difficulties of thermodynamics itself. Even under extreme conditions, an oscillatory regime is initiated which is specific to nonlinear irreversible processes. It is claimed that processes in living systems are regulated by many control systems preventing the organism from entering the nonlinear regions (Lamprecht and Zotin, 1983).

The existence of elaborate control mechanisms for the various biochemical processes inside and within living cells is responsible for the coherent behavior observed in its spatio-temporal organization. Most of the molecular processes controlling different cellular functions are regulated by a few common phenomena, such as sequential reactions with rate-limiting steps, competition for commons sites, allosteric changes, end-product inhibition, and so on.

Stability and sensitivity are both necessary properties of living systems and these are achieved through positive and negative feedback pathways. Study of a three-step reaction scheme (Sinha and Ramaswamy, 1986) with end-product inhibition and allosteric activation coupled to end-product inhibition has shown the exhibition of a variety of behaviors under different conditions. Such behaviors include a steady state, a simple limit cycle, complex oscillations, and period bifurcations leading to chaos. In complex systems, one also observes the existence of two distinct chaotic regions under the variation of a single parameter. Practically all nonlinear oscillators exposed to periodic perturbations display chaotic oscillations for certain values of the controlling parameters. The return to periodic behavior from chaotic behavior through reverse bifurcations is also observed.

The normal living cell is a well-regulated functioning unit. Its metabolic reactions proceed very smoothly at a rate which is remarkably adapted to meet the demands made. The cell can be regarded as the vital unit of organisms and the anatomic and physiologic substrate of biological phenomena. A single living cell has about 3,000 enzymes to affect and regulate its chemical reactions. The enzymes are not distributed randomly within the cells but are located in various compartments. The significance of the high degree of microscopic organization observed in living systems together with the implication this should have on microscopic compartmentalization is probably not fully understood. There is nothing in the chemical structure of the enzyme that can account for its enormous catalytic power.

Enzymatic activity of a cell varies and adapts to such stimuli as increased substrate concentration. The enzymes are frequently disposed of in an orderly fashion within the macromolecular framework of the cell and the cell organoids, which form a multienzyme system. The specificity of an enzyme action has often been modeled on a mechanical "lock and key" principle. This still leaves many features unexplained. It is still necessary to discover how an enzyme (lock) can find the substrate (key) and fit into it without jamming. Undoubtedly, these represent an important means of functional recognition needed for regulation, giving a dynamic quality of control in the differential rates of synthesis and decomposition (Smith, 1988). The increase in complexity in biochemical networks is necessary for the coordination of different activities in the cell.

An enzyme-substrate system can be considered as an amplifier, if one regards the input signal as the amount of enzyme present and the output signals as the amount of reaction product formed per unit time. The characteristics of the amplifier are determined by the feedback path. Any fault which reduces the amount of negative feedback will make the gain tend towards the open-loop gain. Then either the control system will fail because it saturates in the "on" or "off" state, or if the necessary gain-phase criteria are satisfied, it will oscillate; it can do both. In either case, it will cease to exert a controlling function.

Enzymatic inhibition and activation by specific chemical constituents may well exist as a mechanism for inherent regulation of cellular metabolism and function. The mass-law conservation considerations are probably of utmost importance as they pertain to competition for substrate and coenzyme. One of the factors involved in regulation is the concentration of various enzymes present. There appears to exist a qualitative one-to-one relation between particular genes and particular enzymes, whereby genetic determinants are manifested. A particular genetic defect may become evident by the absence of a particular enzyme. Such defects are called metabolic errors, some of which occur naturally, while others are introduced artificially. These aspects may account for regulation of metabolic control qualitatively but not quantitatively.

PERIODIC ENZYME REACTIONS

We now consider a model of enzyme reactions in which the enzyme molecules are arranged spatially, and when excited they interact with the substrate molecules (Frohlich, 1988). The reactions result in the destruction of substrate molecules and the liberation of some energy. Some of the liberated energy is utilized for the activation of other enzymes. Activated molecules have finite lifetimes and they decay back to nonactivated states. Substrate molecules are coherently excited, and attract other substrate molecules into the region. The activation-deactivation of the enzymes gives rise to a corresponding activation-deactivation of the correlated electric dipole, and hence to an electric signal.

If N is the number of nonactivated enzyme molecules, and is very large compared with the number, $N^\star$, which is the number of activated enzymes, then one may regard N as time independent. The rate of change of N is proportional to $N^\star NS$, where S is the number of substrates, a triple collision with the reaction rate A. If B is the rate constant for the spontaneous decay,

$$\{dN^\star/dt\} = AN^\star NS - BN^\star$$
$$\{dS/dt\} = -AN^\star NS + CS \tag{10.5}$$

where C is the rate constant for the attraction of substrates.

These equations have an implicit solution,

$$N^{\star C}S^B = de^{AN(N^\star+S)} \tag{10.6}$$

where d is a constant. In addition, a time-independent solution $N_o^\star$, S_o exists.

$$N_o^\star = (C/AZ); \qquad S_o = B/(AZ) \tag{10.7}$$

Thus, we have

$$n = N^\star - N_o^\star; \qquad s = S - S_o \tag{10.8}$$

These satisfy the Lotka-Volterra equations

$$\{dn/dt\} = Cs + AN + n$$
$$\{ds/dt\} = -Cn - ANsn \tag{10.9}$$

which have time-periodic solutions. Thus, the system does not evolve into a time-independent state when $n = 0$, $s = 0$.

If s and n are so small such that the product sn can be neglected, the solutions are

$$n = D(C)^{(1/2)} \cos wt$$

$$s = D(B)^{(1/2)} \sin wt \tag{10.10}$$

$$w^2 = BC$$

Thus, one has a periodic reaction system, whose frequency is determined by the rate constants for recombination and attraction. As a result, electric field oscillations with the same frequency are generated at each enzyme site. The spatial arrangements of these sites determine to what extent signals are observed at a distance. Frohlich (1988) proceeds to show that a limit cycle exists and that at a later time oscillations approach a state in which energy can be stored which is also stable against certain perturbations. An appropriately small perturbation may, however, cause the liberation of stored energy.

In a cyclic sequence of reactions, it may be argued that the concentrations of the cycled substrate would exert an effect on the overall rates of the processes. The kinetics of such a cycle are dynamically more complicated than the simple two-step cycle of an enzyme-catalyzed reaction. If such reactions are under stationary-state conditions, the side reactions generating or removing these substrates exert a degree of control to the sequence of reactions. It is of interest to question the type of control that one may associate with the structurally arrayed enzymes. It is conceivable that small alterations in the structural relations at each sequential step would have a profound overall influence on rate.

Therefore, enzymes are more than highly efficient and specific catalysts. The activity of an enzyme may depend critically on particular molecules present in its environment. While most biochemical processes are dominated by proteins (enzymes), the control of protein synthesis itself is dominated by the nucleic acids. A consideration of typical free energy diagrams for enzymatic processes in which the free energy of the system is plotted against the reaction coordinate indicates that enzymes are equilibrium thermodynamic machines which may exploit the energy of ligand binding to lower the energy barrier separating the enzyme-substrate and enzyme-product complexes (Fersht, 1985). While enzymes transduce free energy in the sense that they attain states with higher free energy than the ground state, their macroscopic effect is to catalyze the approach of the reaction to equilibrium. The constraints imposed by the openness of the system to the external environment impose new variations, distinct from those in closed systems, on chemical reactions and concentration behavior in the stationary states.

In living systems, the property of stability is necessary to maintain and conserve energy in cellular economy. The property of desensitization to external perturbation is achieved through negative feedback. Feedback activation plays an important part in amplification of the switching on and off and rapid response processes. An optimal combination of stability and sensitivity is achieved in living systems through the combination of positive and negative feedback processes. This kind of organization in biochemical networks can involve both genetic (end-product repression) and metabolic enzyme activation processes, since both are independent of cellular functions.

SYSTEMIC FEEDBACK

Feedback occurs when a process acts upon itself kinetically and in a closed chain of action results in self-enhancement in the case of positive feedback, and self-inhibition in the case of negative feedback. In a reaction of the kind

$$A + \underset{\underset{\text{\tiny \textemdash\textemdash\textemdash\textemdash\textemdash}}{\uparrow}}{X} \xrightarrow{\ k\ } 2X$$

evidently the concentration of the reactant X is altered and such a kind of feedback is called a *nonsystemic feedback*. Most autocatalysis reactions are of the nonsystemic feedback kind. On the other hand, if one has a reaction of the kind

$$A + X \xrightarrow{\ k \twoheadleftarrow\text{\tiny\textemdash\textemdash}\ } P$$

and the feedback is such that the rate constant k is affected, it is then called a *systemic feedback* (Franck, 1979). Most feedback systems in physicochemical systems act not upon the input but instead upon the transmission system.

Systemic feedback is much more frequent in physicochemical systems than nonsystemic feedback. In all cases, where the driving forces of the chemical reactions or the transport processes influence the rate constant or permeability of their own processes, systemic feedback is said to occur. The rate constants of chemical reactions are dependent on temperature. Thus, in nonisothermal systems, all exothermic reactions are subject to thermal feedback. In isothermal systems, the rate constants may also depend on other parameters, such as concentrations of their own products or reactants, on dielectric coefficients, or on electric potentials. The rate constant is related to the activation energy of the reaction. A reactant or a product can influence the activation energy of its own chemical reaction through effects of primary salt effect, or solvent effects and other cooperative factors. Systemic feedback is always nonstoichiometric. This favors oscillations in such systems. The evident restriction of some chemical oscillators to stoichiometric reactions is quite strong and is limiting. Thus, oscillations are more frequently observed in systems with systemic feedback. The temporal behavior of positive and negative feedback is essential for the occurrence and the temporal pattern of physicochemical oscillations.

Consider the formation and the consumption of a species X by reactions of the first order. A plot of $\{dX/dt\}$ versus X for this system (a phase-plane plot) gives two straight lines which intersect at the steady-state value of X. Their slopes correspond to the relevant rate constants. Self-coupling alters the slope of the kinetic characteristic line belonging to the feedback in question. In the case of positive feedback, it is getting steeper with increasing X and flatter in the case of negative feedback. If the positive feedback is strong, three stationary states may occur. The formation and consumption of X exhibits bistability and instability. Therefore, positive feedback destabilizes the system. This is the intrinsic cause of most instability in autocatalytic systems.

On the other hand, negative feedback has a stabilizing effect and generates one stable state only. Systems containing negative feedback exhibit *overshoot phenomena,* which are a means of recovery as a consequence of the stabilizing activity of the negative feedback. If the dynamic characteristic of a system with positive feedback is nonmonotic, then a threshold pattern of trigger processes occurs.

Franck (1979) defines the *principle of antagonistic feedback of physicochemical oscillators* as oscillations which occur as a consequence of an antagonistic interaction of a relatively fast-acting positive feedback of a destabilizing tendency and a slower-acting negative feedback of a stabilizing recovering tendency. The occurrence of oscillations depends not only upon the presence of both kinds of feedback but also upon the correct apportionment of the time parameters of the feedback loops involved.

Consider the temperature of a living system. The usual working temperature of about 300 K is a relatively high value. At infinite volume, an ordered system permits the concept of zero-degree Kelvin. A nonzero temperature would be obtained only by introducing thermal fluctuations. A temperature decrease does not improve the ordering of the system, as one would expect by the fact that fluctuations also decrease. In living systems, the dynamical order disappears when the temperature falls below room temperature.

FLOW PROCESSES ───

The ability of nearly every living organism to maintain a constant internal cellular composition of a series of important inorganic ionic species at concentrations greatly differing from those in the surrounding medium is one of the puzzling features of regulation and control in biology. The cells are able to maintain these constant ionic concentrations, even though the enclosing membranes are permeable to the ions in question. Transport of a substance across a biological membrane is called *passive* if the process can be accounted for by means of ordinary physical reasons. Fluxes which are caused by a concentration gradient, an electric potential gradient, or any combination of these are called passive transport. *Active* transport processes may, therefore, be regarded as those processes which cannot be explained as a result of these forces but involve the participation of some energy-yielding chemical reaction. The conclusion that active transport processes are involved in many cases of transport when ion accumulation and extrusion occur against their (normally accepted definition of) electrochemical occur against their (normally accepted definition of) electrochemical potential gradient is beyond question (Vaidhyanathan, 1977); that the required energy is derived from metabolism has been unequivocally shown. But how and in what form the energy is transferred to the actively transported ion in question is still unknown.

Active transport of ions moves electric charge as well as creating a concentration gradient. The Na^+-K^- pump has been studied for more than half a century. The energy source has been identified as ATP, the concept of a strictly stoichiometric coupled transport has been established, and a suitable ATPase activity has been discovered in fragments of cell membranes. Apparently proteins that couple ion transport to a source of free energy have arisen several times in evolution. Despite the long history of sophisticated experimental observations, one still does not understand, how ions are picked up on one side of the membrane and deposited on the other side.

Barring the existence of a mysterious vital force operative in biological membranes, if physiological systems consist of molecules and ions with corresponding intermolecular interactions over the distance between them, assumed adequately explainable on the basis of our current knowledge, then the concept of active transport boils down to these facts. A biological membrane is able to sustain a large difference in concentration of a specified species in spite of the fact that experiments (performed possibly under different conditions) have shown that this membrane is permeable for the substance in question. The permeability property of an inhomogeneous membrane barrier is a complex function of concentrations, fluxes, partial frictional coefficients, electric potential profile, and chemical reactions, apart from structural contributions and the state of the system. All computations of the so-called general chemical potential difference of the species in question across the two sides of the membrane under active transport conditions are neither complete nor satisfactory.

The influence of chemical reactions on fluxes of species participating in such reactions, is described globally, by the equation of continuity, (4.8), and is generally described by reaction diffusion equations of the kind (6.86). At the molecular level, the flux of a specified species in the presence of a chemical reaction is different from the value in the absence of such chemical reaction, due to two separate reasons: 1) The concentration gradients are now different and; 2) the resistance that the medium profers for its flow is also different in the presence of chemical reactions, from the value of resistance in the absence of such chemical reaction.

In a qualitative description of the phenomenon of active transport, one is forced to assume an asymmetrical property of the membrane system to overcome the constraints of Curie's theorem, including asymmetrical properties of the transport coefficients. Essentially this implies that the mobility and its inverse, the frictional coefficient, is a function of the position variable in the membrane phase. The influence of the chemical reaction on the flux of a species not participating in such a chemical reaction can occur through coupling effects of the partial frictional coefficients, as suggested by molecular theory (Vaidhyanathan, 1971). The molecular theory suggests that the chemical reaction rate profile affects the concentration distributions, thereby affecting the concentration gradients as well as the intermolecular interactions that diffusing particles experience during transit across the membrane system. Therefore, the influence of the chemical reactions can be described to occur through alteration of the local forces and alteration of the local resistance that molecules experience. The usually employed assumption that matter flux is proportional to the difference in concentration, however convenient, is not aesthetically satisfactory or correct.

If one accepts the limited validity of Fick's law and includes the coupling influence of fluxes of one kind of species on the flux of a different kind, and recognizes that the diffusion coefficient cannot be regarded as position independent, one has a reasonable basis for the construction of a quantitative theory of the influence of the chemical reaction on the fluxes of specified species across membranes.

The direction of overall (net) flux and the direction of overall (net) force on specified species of a system undergoing transport should be the same. Thus, the central problem of understanding the active transport mechanism is the proper evaluation of the magnitude and direction of the net force to which a molecule of a particular species is subjected to and the direction of its net flux. The influence of the chemical reaction on the fluxes in an inhomogeneous anisotropic medium is complicated both conceptually and analytically. The primary influence of a chemical reaction is to alter the concentration profile that might have been present in the absence of such chemical reactions, thereby modifying any gradients responsible for such fluxes. A secondary influence can occur through the alteration of the resistance that the moving species encounters. Thus, active transport, visualized to arise due to the utilization of metabolic energy by some ill-defined pumping mechanism, basically boils down to the correct computation of the magnitude and direction of the forces and fluxes involved.

Mathematically, a reaction–diffusion system is obtained by adding some diffusion terms to a set of ordinary differential equations which are first order in time. This set of equations is appropriate for studying various aspects of chemically reacting and diffusing systems. In general, one can state that the diffusion processes tend to drift the system toward a more homogeneous state, while the chemical reactions tend to lead to more inhomogeneity. Thus, the tendencies of these two processes are in mutually opposite directions, and there exists an inherent competition between the two processes.

Similarly, the Donnan membrane potential is the result of the two competing tendencies of the system, one to maintain electroneutrality, and the other to maintain equality of chemical potential in all locations at equilibrium of all permeating species. One can speculate that the gating mechanism, exhibited by ion channels, essentially arises due to some kind of a linear accelerator effect. An incoming charged ionic species polarizes the molecular segments of the walls of the ionic channels, creating an opposite charge. This leads to the creation of additional attraction, resulting in further acceleration of incoming ions, enabling them to zip through. But as more and more ions transit, such polarization becomes cumulative, especially when the dipoles cannot relax back to their equilibrium positions quickly due to steric hindrance. The closing of the ion channels can be suspected to occur when excessive charge of the opposite kind is built up in the walls, so that the walls become extremely attractive and become like glue, adhering to incoming charges, thus preventing them from escape to the other side. This is similar to the difference in approach of a proton near an electron cloud,

---

† Often one finds discussions of the responses of the gating particles in channels to externally applied electric potential differences. These are usually of the order of millivolts. At a distance of about 10 angstroms from a unit charge, monovalent ion-like sodium, when the dielectric coefficient, ε, is about 80 (the dielectric coefficient is more likely to be of the order of unity at these distances), the electric field intensity $\{E = e^2/\varepsilon r^2\}$ caused by this charge can be computed to equal 0.18 million volts per cm. Thus, the electric potential at a distance of 10 angstroms from the approaching univalent ion is the product of 18 millivolts and the ratio $\{80/(\text{the value of the dielectric coefficient in the intervening space})\}$. For passage, ions should approach the channel entrance much closer than 10 angstroms. The dielectric coefficient is more likely to be of the order of unity than 80, and the value of this electric potential is evidently higher at closer proximity. For multivalent ions, this potential should be multiplied by the square of the valence charge number, namely, 4 for divalent ions. The question naturally arises as to when such charges transit the channel: Does the gating process respond to such high potentials in the proximity of transiting charges, or does it respond to externally applied electric potentials which are evidently much smaller in magnitudes?

leading to acceleration, and the immersing action of a proton in a sea of electrons. The further opening of the same channel is accomplished by the relaxation of dipoles to their original state, given sufficient time. Such a principle of adding acceleration to a man-made satellite is illustrated by Galileo's itinerary to Jupiter, where an assist from Venus enables one to impose additional acceleration on the satellite.

Coulombic interactions occur over a long range. Coupled with the fact that the dielectric coefficient of lipid bilayer membranes is of the order of 2, the collective behavior of charged species with respect to the relative motion of each other in biological systems becomes very important. Coulombic interactions also introduce both collective behavior (through the Poisson equation) and the nonlinear properties of relevant differential equations. Since partial frictional coefficients arise from interparticle interactions, they can assume either sign, due to like and unlike charges interactions. Thus, it is in principle possible for the frictional coefficient to alter its magnitude as well as its sign in an inhomogeneous medium depending on the nature of the surrounding molecules and the kind of interactions. Therefore, permeability of a specific species across a membrane can vary in magnitude from system to system. A membrane, which has been permeable to a specific species under a given set of circumstances, can become more permeable or less permeable, or even become impermeable to the same species under a different set of circumstances. This is possibly the plausible basic mechanism by which the transport of species in biological membranes is regulated.

Although contribution to entropy production arises due to the occurrence of chemical reactions, it is usually a poor approximation to write reaction fluxes as a linear homogeneous function of the forces. Apart from the fact that the difference in free energies is a scalar, the reaction flux (which is also a scalar) is never proportional to the Gibbs free energy change for the chemical reaction, except when the reacting system is extremely close to equilibrium.

In the active transport process, work is done for the transport of molecules against their concentration gradients at the expense of energy derived from metabolism. Since nothing is known about how the energy is transferred to an actively transported ion or molecular species, most metabolic studies of transport processes are concerned with the comparison of the work involved in transport with the total energy production. Estimation of the minimum energy requirements of active transport will often be invalidated by the fact that exchange diffusion processes may participate in the movement of molecules to an unknown extent. Although the problem of active transport has been observed to exist for at least five decades, not many serious theories exist to explain the phenomena.

In order to appraise and advance our knowledge on regulation and control in biology, it is imperative that one must review the information available from biochemistry, physical properties of water and electrolytes, and an overall appraisal of various physicochemical processes occurring in the system. It has been proposed that life is made up of three basic components: matter, energy, and information (Fong, 1973). Matter and energy are conserved and can be transformed. Both creation and dissipation of information are understood from the theory of fluctuations and the second law of thermodynamics. Information can be transformed but is not conserved. Iberall and McCulloch (1970) define living systems from a cybernetic point of view as any compact system containing an order and distributon of sustaining nonlinear limit–cycle oscillators and a related system of algorithmic guide mechanisms. Living systems are also viewed as capable of regulating their interior conditions for a considerable range of ambient environmental conditions so as to permit their own satisfactory operation. The essential characteristic of living systems is deemed to be their marginal instability.

The living system, in commerce and trade with its surroundings, tends to increase itself, as an eddy in a stream of energy tends to grow. This propensity for growth furnishes opportunity under factors of evolution for a continual production of modified patterns of the eddy. The pattern evolves in some of them in an increasing complexity. Open systems, like whirlpools, are typically very complicated. They owe their existence to being persistent, not for being in a state of balance in an effectively unchanging world, but to being in a compensating state of imbalance. Organisms are not particularly good at surviving. If organisms are prevalent, it is mainly because they multiply. It is their ability to reproduce that is a peculiar cause for their prevalence. It is in this faculty that the primary biological function exists.

The conclusion that one arrives at from the study of living systems is that populations of coupled cells may give rise to coherent behavior extending beyond the level of a single cell. These phenomena reflect the numerous regulatory and coordination processes exerted in the system. For this reason they are associated with the network behavior of the system. The activity of the central nervous system and the immune response belong to this category.

In the control of growth in multicellular biological systems, long–range interactions and long–range correlations are required. The great sensitivity of biological systems for various external perturbing factors, like weak electromagnetic radiations, is phenomenal. The electric fields of the order of 10^5 volts per cm, prevalent near the biomembranes, can lead to structural changes of polyvalent biopolymers, resulting in various nonlinear effects. Such high fields also cause polarization. Poisson's equation, relating the local charge density to gradients of electric potential and dielectric coefficient, essentially represents mathematically the many-body problem. The collective behavior of charged sites of a protein in response to external electric stimulus is thus guaranteed by the long-range nature of coulombic interactions.

Colinvaux (1978) argues against the validity of the thesis that complexity leads to stability. According to him, a measure of species diversity must also be a measure of complexity. The instability of the arctic animal population obviously has some relation with the highly unstable hostile climate.

In order to describe the essential features of life or living systems, such as replication, more information is needed. By cycles of chemical reactions, organisms appear to be able to maintain their coherence. A new structure is asserted to be always the result of an instability in the dissipative structure theory. Del Giudice *et al.* (1988) assume that a living system is the final step of a dynamical evolution which originates from the basic interactions within a set of electric dipoles. An electric dipole is just a physical schematization of a biomolecule. It is also not easy to define the minimum energy state out of the thermodynamic equilibrium.

REFERENCES

ANDREWS, J.G. and R.R. McLONE, *Mathematical Modeling*, London: Butterworths (1976).

COLINVAUX, R., *Why Big Fierce Animals Are Rare?* Princeton: Princeton Science Library, 1978.

DEL GIUDICE, E., S. DOGLIA, M. MILANI, and G. VITIELO, in *Biological Coherence and Response to External Stimuli*, p. 48, ed. H. Frohlich. Berlin: Springer-Verlag, 1988.

FERSHT, A.R., *Enzyme Structure and Mechanism*. San Francisco: Freeman, 1985.

FONG, P., in *Biogenesis Evolution Homeostasis*, p. 93, ed. A. Locker. Berlin: Springer-Verlag, 1973.

FRANCK, U.F., Plenary Lecture of a Discussion Meeting held by Deutsche Bunsengesellschaft fur Physikalische Chemie. Aachen, 1979.

FROHLICH, H., *Biological Coherence and Response to External Stimuli*. New York: Springer-Verlag, 1988.

HILL, T.L., *Free Energy Transduction in Biology*. New York: Academic Press, 1977.

IBERALL, A.S., and W.S. McCULLOCH, in *Technical and Biological Problems of Control: A Cybernetic Point of View*, p. 39, eds. A.S. Iberall and J.B. Roswick. Washington, D.C.: International Federation of Automatic Control, 1970.

KAISER, F., in *Biological Coherence and Response to External Stimuli*, p. 25, ed. H. Frohlich. New York: Springer-Verlag, 1988.

KREUZER, H.J., *Nonequilibrium Thermodynamics and Its Statistical Foundations*. London: Oxford University Press, 1981.

LAMPRECHT, I., and A.I. ZOTIN, *Thermodynamics and Kinetics of Biological Processes*. Berlin: de Gruyter, 1983.

SINHA, S., and R. RAMASWAMY, in *Chaos in Biological Systems*, p. 59, eds. Degn, Holden, and Olsen. New York: Plenum Press, 1986.

SMITH, C.W., in *Biological Coherence and Response to External Stimuli*, ed. H. Frohlich. New York: Springer-Verlag, 1988.

VAIDHYANATHAN, V.S., *J. Theor. Biology*, vol. 7, (1964), p. 334.

VAIDHYANATHAN, V.S., *J. Theor. Biology*, vol. 31, (1971), p. 53.

VAIDHYANATHAN, V.S., in *Topics in Bioelectrochemistry and Bioenergetics*, vol. 1, p. 287, ed. G. Milazzo. London: John Wiley and Sons, 1977.

ZOTIN, A., *Thermodynamic Aspects of Developmental Biology*, Basel: Karger, (1972).

Author Index

Subject Index